Alexander Kanitz

Post-Transcriptional Gene Regulation

Alexander Kanitz

Post-Transcriptional Gene Regulation

Tools and Strategies for the Unraveling of Post-Transcriptional Gene Regulatory Networks

Südwestdeutscher Verlag für Hochschulschriften

Impressum / Imprint

Bibliografische Information der Deutschen Nationalbibliothek: Die Deutsche Nationalbibliothek verzeichnet diese Publikation in der Deutschen Nationalbibliografie; detaillierte bibliografische Daten sind im Internet über http://dnb.d-nb.de abrufbar.

Alle in diesem Buch genannten Marken und Produktnamen unterliegen warenzeichen-, marken- oder patentrechtlichem Schutz bzw. sind Warenzeichen oder eingetragene Warenzeichen der jeweiligen Inhaber. Die Wiedergabe von Marken, Produktnamen, Gebrauchsnamen, Handelsnamen, Warenbezeichnungen u.s.w. in diesem Werk berechtigt auch ohne besondere Kennzeichnung nicht zu der Annahme, dass solche Namen im Sinne der Warenzeichen- und Markenschutzgesetzgebung als frei zu betrachten wären und daher von jedermann benutzt werden dürften.

Bibliographic information published by the Deutsche Nationalbibliothek: The Deutsche Nationalbibliothek lists this publication in the Deutsche Nationalbibliografie; detailed bibliographic data are available in the Internet at http://dnb.d-nb.de.

Any brand names and product names mentioned in this book are subject to trademark, brand or patent protection and are trademarks or registered trademarks of their respective holders. The use of brand names, product names, common names, trade names, product descriptions etc. even without a particular marking in this works is in no way to be construed to mean that such names may be regarded as unrestricted in respect of trademark and brand protection legislation and could thus be used by anyone.

Coverbild / Cover image: www.ingimage.com

Verlag / Publisher:
Südwestdeutscher Verlag für Hochschulschriften
ist ein Imprint der / is a trademark of
AV Akademikerverlag GmbH & Co. KG
Heinrich-Böcking-Str. 6-8, 66121 Saarbrücken, Deutschland / Germany
Email: info@svh-verlag.de

Herstellung: siehe letzte Seite /
Printed at: see last page
ISBN: 978-3-8381-3670-7

Zugl. / Approved by: Zürich, ETH, Diss., 2011

Copyright © 2013 AV Akademikerverlag GmbH & Co. KG
Alle Rechte vorbehalten. / All rights reserved. Saarbrücken 2013

Table of Contents

1 SUMMARY _____ 4
1.1 Summary _____ 4
1.2 Zusammenfassung _____ 7

2 INTRODUCTION _____ 10
2.1 Key principles and players of post-transcriptional gene regulatory processes _____ 11
2.1.1 The fate of eukaryotic messenger RNAs _____ 11
2.1.2 *Cis*-regulatory elements _____ 13
2.1.3 *Trans*-acting factors _____ 14
2.1.3.1 RNA-binding proteins _____ 14
2.1.3.2 MicroRNAs _____ 15
2.1.4 Ribonucleoprotein complexes and the RNA regulon theory _____ 17
2.2 Post-transcriptional gene regulatory networks _____ 20
2.2.1 Basic network motifs _____ 21
2.2.1.1 Multiple output network motifs _____ 21
2.2.1.2 Multiple input network motifs _____ 23
2.2.2 Autoregulatory, two- and multicomponent loops _____ 24
2.2.3 Composite gene regulatory networks _____ 27
2.2.3.1 RNA-binding proteins versus transcription factors _____ 27
2.2.3.2 RNA-binding proteins versus microRNAs _____ 28
2.3 Ribonomics methodologies for the systematic identification of basic post-transcriptional gene regulatory network motifs _____ 30
2.3.1 Top-down approaches: RIP-Chip and related methods _____ 31
2.3.2 Bottom-up: RNA affinity chromatography and related methods _____ 33
2.3.2.1 Direct RNA affinity chromatography _____ 33
2.3.2.2 Purification methods based on antisense hybridization _____ 34
2.3.2.3 Aptamer-based purification methods _____ 35
2.3.2.4 Indirect protein- and peptide-based purification methods _____ 36
2.3.2.5 Bifunctional RNA tag systems _____ 37
2.4 Combinatorial control of cancer-related messages _____ 38
2.4.1 Combinatorial control of the angiogenesis factor vascular endothelial growth factor A _____ 38
2.4.2 Combinatorial control of the tumor suppressor CDKN1B/p27 _____ 42

3 AIMS AND OUTLINE OF THE THESIS _____ 46

4 IDENTIFICATION OF NEW POST-TRANSCRIPTIONAL REGULATORS OF VASCULAR ENDOTHELIAL GROWTH FACTOR A EXPRESSION _____ 48
4.1 Introduction _____ 48
4.2 Results _____ 53
4.2.1 The VEGFA 3'-untranslated region contains canonical Pum consensus motifs _____ 53
4.2.2 VEGFA is a putative target of microRNA 361-5p _____ 57
4.2.3 MicroRNA 361-5p and Pum1/2 may target other angiogenesis-related transcripts _____ 59
4.2.4 Generation and characterization of stable Pum1/2 overexpression cell lines _____ 62
4.2.5 Transfection of small RNAs _____ 65
4.2.6 The putative Pum and microRNA 361-5p recognition elements in the VEGFA 3'-UTR possess regulatory potential _____ 66
4.2.7 Pum1, Pum2 and microRNA 361-5p repress the expression of VEGFA 3'-UTR reporters _____ 69
4.2.8 The repressive effects of Pum proteins and microRNA 361-5p on VEGFA 3'-UTR reporter activity are additive _____ 71
4.2.9 Endogenous VEGFA expression is regulated by microRNA 361-5p _____ 74
4.2.10 MicroRNA 361-5p is down-regulated in cutaneous squamous cell carcinoma _____ 76
4.3 Discussion _____ 81

4.3.1	The putative Pum and microRNA recognition elements exhibit regulatory potential	81
4.3.2	Combinatorial control of VEGFA expression by microRNA 361-5p and the Pum proteins	83
4.3.3	The regulation of VEGFA expression by microRNA 361-5p and the Pum proteins may be dependent on each other	85
4.3.4	The influence of microRNA 361-5p and Pum1 on VEGFA secretion rates	87
4.3.5	A potential role of microRNA 361-5p and Pum proteins in cancer development and progression	88
4.3.6	Bioinformatics analyses suggest common functions of microRNA 361-5p and Pum proteins beyond the regulation of VEGFA expression	90
4.3.7	Conclusion	91

4.4 Materials and Methods _____ 91
4.4.1	Ethics statement	91
4.4.2	Plasmids	91
4.4.3	Cell culture and tissue samples	93
4.4.4	MicroRNA target gene prediction and pathway analysis	95
4.4.5	Immunoblot analysis	96
4.4.6	Flow cytometry	97
4.4.7	Quantitative reverse transcription PCR	97
4.4.8	Immunocytochemistry	98
4.4.9	Luciferase reporter assays	99
4.4.10	Enzyme-linked immunosorbent assay	100

4.5 Contributions _____ 100

5 A NOVEL RNA TANDEM AFFINITY TAG FOR THE PURIFICATION OF RIBONUCLEOPROTEIN PARTICLES _____ 102

5.1 Introduction _____ 102

5.2 Results _____ 104
5.2.1	Aptamer selection	104
5.2.2	Oligonucleotide selection	106
5.2.3	Arrangement of the HAMMER tandem affinity tag system	108
5.2.4	Purification strategy	110
5.2.5	Plasmid generation	112
5.2.6	Secondary structures of HAMMER-tagged RNAs are largely unaffected	114
5.2.7	HAMMER-tagged RNAs are expressed in transiently transfected cells	117
5.2.8	Purification of HAMMER-tagged *in vitro* transcripts via hybridization to antisense oligonucleotides	120
5.2.9	Purification of HAMMER-tagged *in vitro* transcripts via the S1 aptamer	123

5.3 Discussion _____ 125
5.3.1	Expression of tagged transcripts	125
5.3.2	Capturing of tagged transcripts by antisense hybridization	126
5.3.3	Elution of transcripts immobilized by hybridization	127
5.3.4	Aptamer-mediated purification of tagged transcripts	128
5.3.5	Reflections on tag folding and insertion	129
5.3.6	Limitations of RNA secondary structure prediction algorithms	130
5.3.7	Conclusion	131

5.4 Materials and Methods _____ 132
5.4.1	Tag design and bioinformatics	132
5.4.2	Plasmids	132
5.4.3	Cell culturing	133
5.4.4	Quantitative reverse transcription PCR	133
5.4.5	Fluorescence microscopy	134
5.4.6	*In vitro* transcription and labeling	135
5.4.7	S1 aptamer purification	136
5.4.8	Oligonucleotide synthesis	136
5.4.9	Preparation of antisense oligonucleotide matrix	138
5.4.10	Purification by antisense oligonucleotide hybridization	138

5.5	Contributions		139
6	**CONCLUDING REMARKS**		**140**
7	**APPENDIX**		**143**
7.1	Nucleotide sequences		143
	7.1.1	Nucleotides used for cloning	143
	7.1.2	Nucleotides used for mutagenesis	143
	7.1.3	Nucleotides used for quantitative reverse transcription PCR (SYBR Green)	144
	7.1.4	Nucleotides for the HAMMER RNA tandem tag	144
7.2	MicroRNA mimics and antisense inhibitors		145
7.3	Commercial quantitative reverse transcription PCR assays		145
7.4	MicroRNAs predicted to target VEGFA		146
7.5	Predicted microRNA 361-5p targets		153
7.6	Gene set enrichment analysis of microRNA 361-5p, Pum1 and Pum2 targets		160
7.7	pTO-HA-Strep-GW-FRT map and sequence		162
8	**BIBLIOGRAPHY**		**164**
9	**ACKNOWLEDGMENTS**		**187**
10	**ABBREVIATIONS**		**189**

1 Summary

1.1 Summary

In order to guarantee survival in a complex and ever-changing environment, cells dispose of efficient and highly dynamic regulatory circuits that continuously interpret the genetic code in a context-dependent manner. This ensures, with remarkable accuracy, that the right components are at the right place at the right time. Consequently, messenger RNAs (mRNAs) – which play a central role in the flow of genetic information by carrying it from the nucleus to the cytoplasm or, ultimately, from DNA to proteins – are subject to particular scrutiny by a cell's regulatory machinery. It is believed that at every instant of an mRNAs life it is decorated by a host of ever-changing '*trans*-acting factors', namely RNA-binding proteins and non-coding RNAs, which assemble with adaptor, scaffolding and effector proteins into dynamic macromolecular ribonucleoprotein (RNP) complexes. RNPs represent the functional units of post-transcriptional gene regulation, and it is their respective compositions that dictate which gene regulatory program is executed for a particular mRNA at a given moment, i.e. whether they are spliced, edited, exported, silenced, translated, transported, stored, or degraded. Whether a protein or non-coding RNAs is part of an RNP is determined by its availability and activity, as well as the nature of the message itself: Next to the protein coding information, each mRNA species contains a unique set of distinct '*cis*-regulatory elements', sequence and/or structural motifs that are recognized by the *trans*-acting factors. To a certain degree, the fate of an mRNA is therefore pre-determined by its underlying 'RNA code'.

RNPs are organized into highly intertwined, decentralized post-transcriptional regulatory networks which constantly exchange information with the intra- and extracellular environment. They receive and integrate external stimuli, interpret and eventually relay them

Summary

in order to coordinate RNP remodeling accordingly. In order to fully understand the logic and logistics of living systems, it is therefore paramount to unravel these intricate networks. Detailed insights into their nature and organization principles should have broad implications for a large number of scientific and technological disciplines, including but not limited to medicine, computer sciences and even socioeconomics. Here we present strategies and tools for the identification of combinatorial control motifs, a fundamental component of such networks.

In a first project, we integrated bioinformatics analyses and experimental evidence to predict, with high confidence, *cis*-regulatory elements for the RNA-binding proteins Pum1 and Pum2, and the non-coding RNA microRNA 361-5p in the message of the angiogenesis factor vascular endothelial growth factor A (VEGFA). Using *in vitro* reporter assays we then demonstrated that the defined sequence elements possess regulatory potential and that elevated levels of the corresponding *trans*-acting factors negatively affect reporter activities in a combinatorial, additive manner. RNA levels of microRNA 361-5p, Pum1 and Pum2 were all reduced in human cutaneous squamous cell carcinoma samples that express elevated levels of VEGFA, suggesting a potential role in tumor development and/or progression. The study represents a valid strategy for the prediction of combinatorial control motifs in an mRNA of interest. It further lays the foundation for future studies that could address the relevance and interplay of these new repressors and their regulatory elements for the regulation of VEGFA expression in more detail, such as *in vivo* and in disease.

In a second project, we present a strategy for studying the composition of RNPs. To this end, we rationally designed a modular tandem tag system that contains an RNA sequence with high affinity for the ligand streptavidin ('streptavidin-aptamer'), as well as an exposed RNA oligonucleotide without strong secondary structure characteristics. The tag can be attached to an RNA of interest and is supposed to allow the purification of the corresponding RNP in a highly specific two-step manner that involves the interaction of the aptamer with immobilized streptavidin on the one hand, and hybridization of the exposed RNA oligonucleotide to an immobilized antisense strand composed of a stable RNA derivative on the other. Due to gentle elution methods, the procedure should be compatible with downstream applications for the identification of its protein and RNA components. So far, we have generated a number of tools and controls for the characterization of the tag system and were able to show in preliminary experiments that a tagged *in vitro* transcript could be enriched by antisense hybridization, but not aptamer-based purification. If the method can be successfully established in future experiments, it should find widespread use for the characterization of RNP composition and plasticity.

1.2 Zusammenfassung

Um das Überleben in einer komplexen und sich ständig verändernden Umgebung zu garantieren, verfügen Zellen über hocheffiziente und dynamische Regelkreise, die den genetischen Kode fortlaufend und situationsabhängig interpretieren, um mit bemerkenswerter Sorgfalt dafür zu sorgen, dass sich die richtigen Bausteine zur rechten Zeit am rechten Ort befinden. Boten-RNAs (mRNAs) spielen eine Schlüsselrolle im Fluss genetischer Information, indem sie diese vom Zellkern in das Zellplasma beziehungsweise von der DNA zum Protein tragen, und unterliegen folglich einer besonderen Aufmerksamkeit durch die Steuermechanismen einer Zelle. Es wird angenommen, dass eine mRNA zu jedem Zeitpunkt ihres Lebens von einer Vielzahl ständig wechselnder „*trans*-aktiver Faktoren" - RNA-bindende Proteine und nicht-kodierende RNAs - gebunden wird, die sich mit Adapter-, Gerüst- und Effektorproteinen zu makromolekularen Ribonukleoproteinkomplexen (RNPs) zusammenlagern. RNPs stellen die funktionellen Einheiten der post-transkriptionellen Genregulation dar, und ihre jeweilige Zusammensetzung bestimmt, welches genregulatorische Programm für welche mRNA zu einem bestimmten Zeitpunkt ausgeführt wird, also ob sie gespleisst, editiert, exportiert, transportiert, gelagert, abgebaut oder aber in ein Protein „übersetzt" wird. Ob ein bestimmtes Protein oder eine nicht-kodierende RNA Teil eines RNPs ist, hängt sowohl von dessen Anwesenheit und Aktivität ab, als auch von der Boten-RNA selbst: Neben dem Bauplan für ein Protein enthält jede mRNA-Spezies nämlich zusätzlich eine einzigartige Kombination verschiedener „*cis*-regulatorischer Elemente" – sequenzbasierte beziehungsweise strukturelle Motive, die von den *trans*-aktiven Faktoren erkannt werden. Zu einem gewissen Grad ist das Schicksal einer mRNA also durch den ihr zugrundeliegenden „RNA Kode" vorgegeben.

RNPs sind in stark verflochtenen, dezentralen sogenannten „post-transkriptionellen regulatorischen Netzwerken" organisiert, die im regen Austausch mit der Umgebung im

Zellinneren und –äusseren stehen. Sie nehmen ständig Reize auf, integrieren und interpretieren diese und leiten sie schliesslich weiter, um so den Umbau der RNPs entsprechend zu koordinieren. Um die Logik und Logistik lebender Systeme zu erfassen, ist es daher unerlässlich, diese hochkomplexen Netzwerke zu entschlüsseln. Detaillierte Einblicke in ihre Beschaffenheit und Organisationsprinzipien könnten von grosser Bedeutung für eine Vielzahl wissenschaftlicher und technologischer Disziplinen sein, etwa der Medizin und der Informatik, aber auch der Erforschung gesellschaftlicher Strukturen. In dieser Arbeit wurden Strategien und Werkzeuge entwickelt und angewandt, um ein grundlegendes Motiv solcher Netzwerke, die „kombinatorische Kontrolle" einer bestimmten Boten-RNA durch verschiedene *trans*-wirkende Faktoren zu untersuchen.

Dazu wurden in einer ersten Studie bioinformatische Analysen mit experimentell-empirischen Erkenntnissen gekoppelt, um so *cis*-regulatorische Elemente für die RNA-bindenden Proteine Pum1 und Pum2 sowie für die nicht-kodierende RNA mikroRNA 361-5p in der Boten-RNA für den Angiogenesefaktor „vascular endothelial growth factor A" (VEGFA) mit hoher Wahrscheinlichkeit vorherzusagen. Mit künstlichen Reportertestverfahren konnten wir dann nachweisen, dass diese in der Tat regulierende Eigenschaften besitzen. Erhöhte Pegel der entsprechenden *trans*-aktiven Faktoren führten ferner zu einer Verminderung der Reporteraktivität in einer kombinatorischen, additiven Weise. Zuletzt konnten wir zeigen, dass die Pegel von mikroRNA 361-5p, Pum1 und Pum2 in menschlichen Plattenepithelkarzinomen reduziert sind und sich konträr zu denen von VEGFA verhalten. Diese Befunde deuten auf eine mögliche Rolle in der Entstehung von Tumoren oder dem Krankheitsverlauf bestimmter Krebsleiden hin. Die Arbeit stellt eine Strategie für die Vorhersage bestimmter Steuerelemente in Boten-RNAs vor. Weiterhin liefert sie Voraussetzungen für künftige Studien, die sich näher mit der Bedeutung und dem Zusammenspiel der identifizierten Repressoren und ihrer Bindestellen für die Regulierung der

Summary

VEGFA-Aktivität beschäftigen könnten, insbesondere bei der Entstehung von Krankheiten. Dies liesse sich durch die Untersuchung von Patienten oder entsprechender Modellorganismen, etwa Mäusen, bewerkstelligen.

In einer zweiten Studie präsentieren wir eine Strategie für die Untersuchung der Zusammensetzung von RNPs, die sich um eine Boten-RNA gebildet haben. Zu diesem Zweck haben wir ein Nukleinsäure-basiertes Erkennungsmerkmal entworfen, welches einerseits aus einer Sequenz mit hoher Affinität zu dem Liganden Streptavidin besteht ('Streptavidin-Aptamer'), und andererseits aus einem exponierten RNA-Oligonukleotid, welches keinerlei besondere Sekundärstrukturmerkmale aufweist. Das Erkennungsmerkmal lässt sich an RNAs anhängen, um so die Aufreinigung der entsprechenden RNPs in einem hochspezifischen, zweistufigen Prozess zu ermöglichen, der die Wechselwirkung mit immobilisertem Strepatvidin beziehungsweise einem Gegenstrang, bestehend aus einem RNA-Derivat, beinhaltet. Ferner ist das Merkmal nach dem Baukastenprinzip aufgebaut, sodass sich einzelne Bauteile mühelos austauschen lassen. Aufgrund schonender Ablöseverfahren der „eingefangenen" RNPs sollte das Verfahren verträglich mit den verfügbaren hochempfindlichen Analysemethoden zur Bestimmung von Proteinen und RNAs sein. Bisher haben wir eine Reihe von Werkzeugen und Kontrollen für die Charakterisierung des Erkennungsmerkmals entwickelt und konnten in ersten Untersuchungen zeigen, dass eine künstlich hergestellte, mit dem Erkennungsmerkmal versehene Boten-RNA mittels Gegenstrang-Hybridisierung spezifisch angereichert werden konnte. Die Aptamer-basierte Aufreinigung hingegen war bisher nicht erfolgreich. Sollte sich diese Methode in künftigen Studien erfolgreich etablieren lassen, ist zu erwarten, dass sie ein breites Anwendungsgebiet für die Untersuchung von RNPs finden wird.

2 Introduction

The ability to establish a high level of order and suspend the harmful effects of entropy is a unique and fascinating quality of living systems. So how does a cell achieve this spectacular feat in a microscopic space that is thronged with millions of molecules of varying sizes, shapes and chemical properties?

In the year 1961 François Jacob and Jacques Monod described the first gene regulatory system, the *Escherichia coli* lac operon (Jacob and Monod, 1961). In the very same year, Sydney Brenner and others identified an "unstable intermediate carrying information from genes to ribosomes for protein synthesis" (Brenner *et al.*, 1961; Gros *et al.*, 1961), yet at the time probably few people had imagined the myriad forms of regulation that these messenger RNA molecules (mRNAs) undergo during their brief, yet eventful – and highly promiscuous – life. Shortly after, the ground work for the deciphering of the genetic code was laid (Crick *et al.*, 1961; Matthaei *et al.*, 1962) and modern molecular biology was born. In the ensuing half century enormous efforts have been devoted to further our understanding of the 'logic of life', gradually introducing scientists to an overwhelmingly complex and intricately intertwined, yet remarkably robust and efficient regulatory circuitry.

This work deals with the post-transcriptional aspects of this circuitry, i.e. the regulatory events that an mRNA undergoes after it is generated. In this chapter, we introduce the underlying concepts, processes and key players of post-transcriptional gene regulation (PTGR; see 2.1), followed by a phenomenological description of post-transcriptional gene regulatory networks (GRNs; see 2.2) and an overview of the methodologies devised for the unraveling of such networks (see 2.3). Finally, we touch upon the relevance of PTGR and its medical implications by summarizing in detail the available literature on the combinatorial control exerted on the messages coding for two proteins with major roles in cancers (see 2.4).

2.1 Key principles and players of post-transcriptional gene regulatory processes

2.1.1 The fate of eukaryotic messenger RNAs

The genome stores the information that is needed to build proteins in defined entities, called genes, in the form of a four-letter DNA polymer. But in order to deliver this information to ribosomes, the cytoplasmic protein production 'factories', a gene's intrinsic sequence code first needs to be transcribed into a mobile carrier, the messenger RNA. For decades, transcriptional regulation of gene expression has been regarded as the dominant force in determining the fate of mRNA transcripts. However, cellular mRNA levels have been demonstrated to be poor indicators of protein abundance (Gygi *et al.*, 1999), indicating that the previous understanding of mRNAs as "blind" carriers of protein-coding information is much too linear. Apparently, the presence of an mRNA molecule in the cytoplasm does not automatically result in the synthesis of the corresponding protein; instead, translation seems to be controlled by a complex logical gate, integrating the availability and state of activity of various molecules in deciding whether an individual transcript is translated or not (Halbeisen *et al.*, 2008; Mansfield and Keene, 2009; Kanitz and Gerber, 2010). It is now clear that regulatory cues are exerted on mRNAs throughout their life, starting during their 'birth' and ending, eventually, with their 'death' – when they are broken up into their individual building blocks. In between, there is a large number of individually regulated steps that ensure that the genetic information is carried in the correct form to where it is required and at the precise time when it is required (Figure 2.1).

Once a eukaryotic mRNA precursor is formed in the nucleus, it first undergoes various processing steps, namely 5'-end processing ('capping'; reviewed in Topisirovic *et al.*, 2011), 3'-end cleavage and 3'-end processing ('polyadenylation'; reviewed in Tian and Graber, 2011). Usually splicing mechanisms further process the mRNA precursor by cutting out long stretches of non-coding information ('introns') that are interspersed between the protein-coding information ('exons'; reviewed in McManus and Graveley, 2011). However, frequently exons are clipped as well, thereby changing the identity of the resulting mRNA, and thus, eventually, of the corresponding protein. The majority of mature eukaryotic mRNAs further contain non-coding sequences on either end of the molecule, the 5'- und 3'-untranslated regions (UTRs), which serve important regulatory purposes, as discussed in the next chapter. Furthermore, introns may exhibit regulatory functions of their own (reviewed in Pyle, 2010). Additional changes to an mRNA's sequence may be made ('editing'; reviewed in Godfried Sie and Kuchka, 2011), before the mRNA is finally prepared for its export to the cytoplasm. The now 'mature' mRNA may be actively localized to a certain

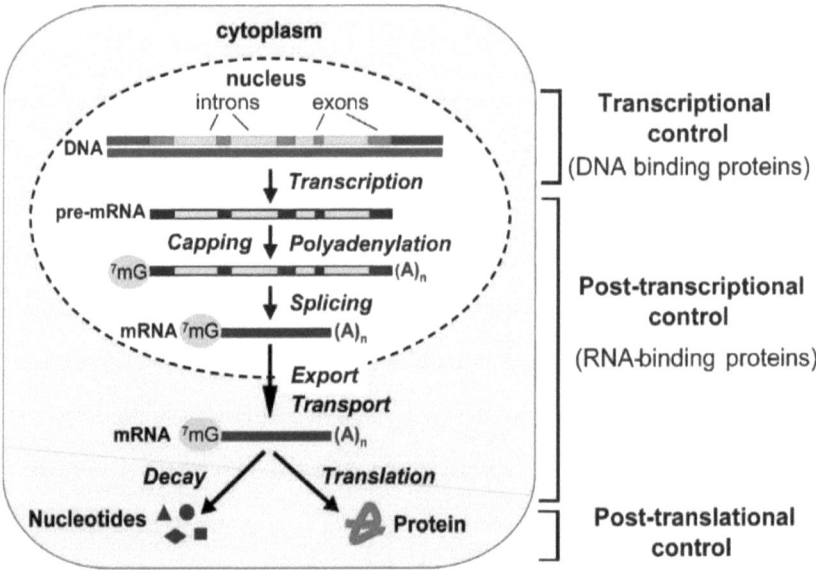

Figure 2.1 Eukaryotic gene expression. The different steps are schematically depicted. See text for details. From Halbeisen *et al.* (2008)

position inside the cell ('RNA localization'; reviewed in Shahbabian and Chartrand, 2011), where it may be stored, translated into protein by ribosomes, or – ultimately – degraded. Regulation of all these different 'programs' of mRNA regulation requires two basic components: Regulators in *trans* (i.e. they act upon other molecules) and recognition elements in *cis* (i.e. they are acted upon by other molecules).

2.1.2 *Cis*-regulatory elements

RNAs are not stiff linear polymers (as frequently depicted in schemes), but rather fold into dynamic three-dimensional structures reminiscent of the folding of proteins. However, due to the smaller number of building blocks and the chemical and structural properties of nucleotides compared to amino acids, the complexity of RNA folding is considerably reduced. *Cis*-regulatory elements are either sequence and/or structural features of an mRNA molecule that are recognized by RBPs, miRNAs and other *trans*-acting factors. While mRNAs often contain *cis*-regulatory elements in their coding sequences, the majority of those features is located in a message's 5'- or 3'-UTRs (reviewed in Mignone et al., 2002; Figure 2.2), probably because the protein-coding information carried by mRNAs puts constraints on RNA folding and the evolution of recognition elements. The usually lower degree of conservation of untranslated regions supports this idea, suggesting that they evolve faster, thus allowing swift (on evolutionary scales) rewiring of regulatory circuits. Moreover, the placement of *cis*-regulatory elements in the untranslated regions has the additional advantage that elements may be distributed more freely, since only a structural, but not a continuous sequence context has to be maintained. While the majority of *cis*-regulatory elements relies on predominantly structural features, some of them can be well represented by degenerate 'consensus sequence motifs' (Figure 2.2).

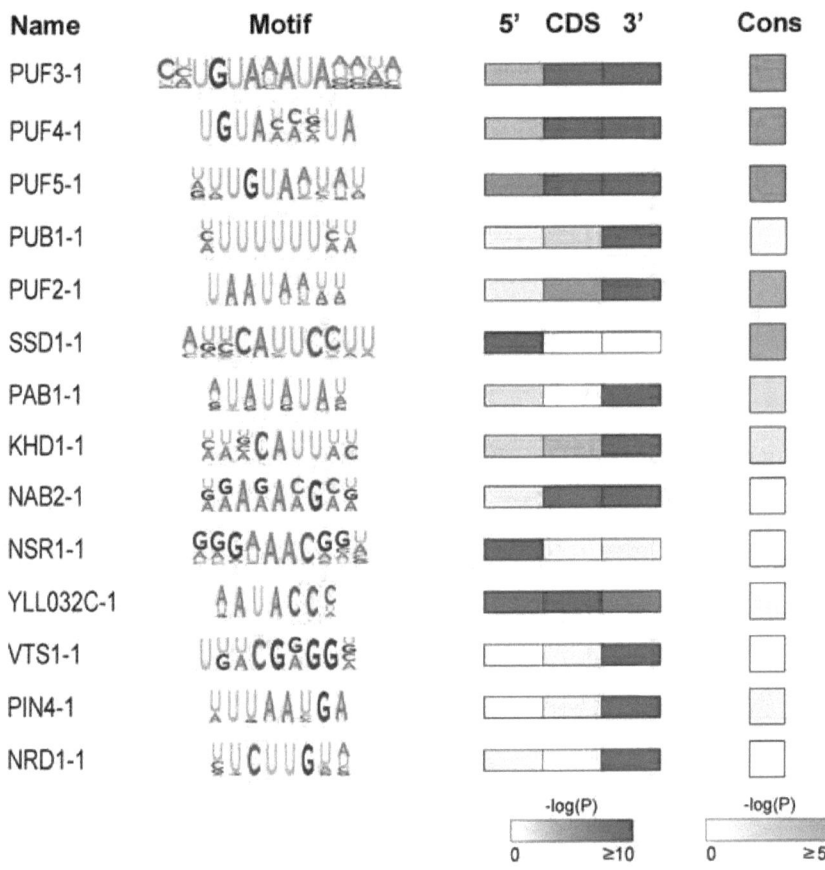

Figure 2.2 Consensus RNA recognition elements of several yeast RNA-binding proteins. The names of the proteins, the experimentally determined consensus motifs, the probabilities that motifs occur in the coding sequence, 5'- or 3'-untranslated regions, and the conservation probability of the motifs are indicated. Adapted from Hogan *et al.* (2008).

2.1.3 *Trans*-acting factors

2.1.3.1 RNA-binding proteins

A considerable fraction of eukaryotic genomes possess RNA-binding activity, with estimations ranging from 8 to 15% of the protein coding genes (Keene, 2001). Approximately 1000-2000 RBPs are encoded in the human genome (Anantharaman *et al.*, 2002). This large number of RNA binding proteins reflects both the ancient role of RNA-dependent regulatory mechanisms, as well as the increased regulatory requirements in eukaryotes, owing to their higher levels of organization through compartmentalization (Keene 2001; Anantharaman *et al.*, 2002). This is particularly true for multicellular organisms, where cell-cell communication adds yet another layer of organization. Consequently, the number of RBPs in higher eukaryotes rivals those of other classes of gene regulators, such as transcription factors (TFs) and kinases. In fact, a recent systematic approach aimed at the identification of novel RBPs in yeast found a high number of proteins with hitherto unappreciated RNA-binding activities, unexpectedly including a number of enzymes, thus suggesting that the real number of RNA-binding proteins may be even larger (Scherrer *et al.*, 2010). RBPs usually contain defined RNA-binding domains through which they interact with *cis*-regulatory elements in the targeted mRNAs. Approximately one hundred distinct RNA-binding domains have been characterized so far, some of which are also able to bind double-stranded RNA (Anantharaman *et al.*, 2002; Lasko, 2003). In contrast to the RBPs with clearly defined roles, such as those required for mRNA processing, there is also a number of 'regulatory' RBPs with less well defined functions that generally stabilize or destabilize bound messages, or inhibit their translation (Shyu *et al.*, 2008).

2.1.3.2 MicroRNAs

Although originally described in 1993 (Lee *et al.*, 1993), the significance of the class of small (~22 nt) non-coding RNAs referred to as microRNAs has only begun to be appreciated with the back-to-back publication of three articles in 2001 (Lagos-Qintana *et al.*, 2001, Lau *et al.*, 2001, Lee and Ambros, 2001). This marked the beginning of an ongoing phase of growing interest in PTGR and the role of non-coding RNAs. MicroRNAs (miRNAs) are endogenously expressed in metazoans, where they exert their repressive function by imperfect antiparallel hybridization to targeted mRNA transcripts. This is mainly mediated via the miRNA's 5'-terminal 'seed' sequence of 6-8 nucleotides with usually perfect or near-perfect complementarity to the target sequence, the miRNA recognition element (MRE; reviewed in Filipowicz *et al.*, 2008). Targets are generally repressed through mechanisms that either lead to enhanced decay or translational inhibition (reviewed in Djuranovic *et al.*, 2011), although some reports demonstrated an activating role under certain conditions (reviewed in Vasudevan, 2011).

Importantly, the functional unit of miRNA-mediated repression is a cytoplasmic ribonucleoprotein complex, termed miRNP (reviewed in Mourelatos *et al.*, 2002), composed of proteins of the Argonaute family of proteins (reviewed in Höck and Meister, 2008), the RNase III endonuclease Dicer, as well as facultative accessory proteins (reviewed in Steitz and Vasudevan, 2009). Based on bioinformatics analyses, it is estimated that miRNAs regulate the majority of all human genes – with each miRNA being able to bind up to hundreds of target mRNAs (Lewis *et al.*, 2005). With more than one thousand miRNAs being encoded in the human genome, they represent a major class of post-transcriptional *trans*-acting factors that are implicated in virtually all biological processes, as well as a wide variety of pathological conditions, particularly cancers (reviewed in Calin and Croce, 2006; Ventura and Jacks, 2009). In some sense, miRNPs can be regarded as RBPs that can be loaded with different target specificities. This is highly advantageous for cells that rely heavily on post-transcriptional control, since the evolution of short nucleotide stretches is considerably easier to achieve than the generation of a RNA-binding domains with different or novel specificities, as is exemplified by the considerably higher number of miRNAs compared to RNA-binding domains (>1000 vs. 100; compare above).

2.1.4 Ribonucleoprotein complexes and the RNA regulon theory

Trans-acting factors usually do not act alone. Instead, it takes the concerted action of various RNA-binding proteins, scaffolding factors and effectors, as well as non-coding RNAs, which assemble into dynamic macromolecular complexes. These structures are referred to as ribonucleoprotein complexes or particles (RNPs) and represent the functional units of PTGR. RNPs assemble on mRNAs as soon as they are transcribed and accompany it throughout their life time while continuously changing their compositions along the way. In fact, it may vary so much that different, relatively stable complexes can be identified at the various stages of an mRNA's life. Moreover, proteins usually do not join or leave an RNP individually, but may be present as pre-formed protein complexes that await an mRNA (or rather the 'previous' RNP), at which point they then 'take over' by displacing factors that are (currently) not needed. In this way, all of the events outlined in chapter 2.1.1 are regulated by RNPs of a more or less defined nature (reviewed in Hieronymus and Silver, 2004; Keene, 2007; Halbeisen *et al.*, 2008; Kanitz and Gerber, 2010). For instance: Spliceosomes regulate the splicing of mRNAs, ribosomes govern the translation into proteins, stress granules may store mRNAs 'for further use', and processing bodies are implicated in their decay (reviewed in Kedersha and Anderson, 2007; Erickson and Lykke-Andersen, 2011; Thomas *et al.*, 2011).

Using ribonomics techniques for the global identification of the RNA targets of RBPs (see 2.3.1), various groups were able to establish that RBPs may bind and regulate subsets of functionally and cytotopically related target mRNAs (Gerber et al., 2004; reviewed in Halbeisen et al., 2008; Morris et al., 2010). This important feature of RNP organization was confirmed in numerous studies conducted in various organisms, suggesting the presence of highly dynamic, transcript- and condition-specific RNPs of a modular architecture, which coordinate post-transcriptional control mechanisms in a combinatorial fashion (Hieronymus and Silver, 2004; Halbeisen *et al.*, 2008). In a series of seminal review articles, Jack Keene first proposed the notion of 'post-transcriptional operons' for such structures (Keene, 2001), in analogy to the bacterial operons, polycystronic transcription units coding for multiple proteins of similar function; and later proposed the existence of 'RNA regulons', higher-level regulatory circuits encompassing multiple post-transcriptional operons for more complex control mechanisms (Keene, 2007). A hypothetical RNA regulon is outlined in Figure 2.3, highlighting the modular architecture and the individual post-transcriptional operons. Figures 2.4 and 2.5 schematically depict the combinatorial and cooperative control concepts of RNA regulons, respectively.

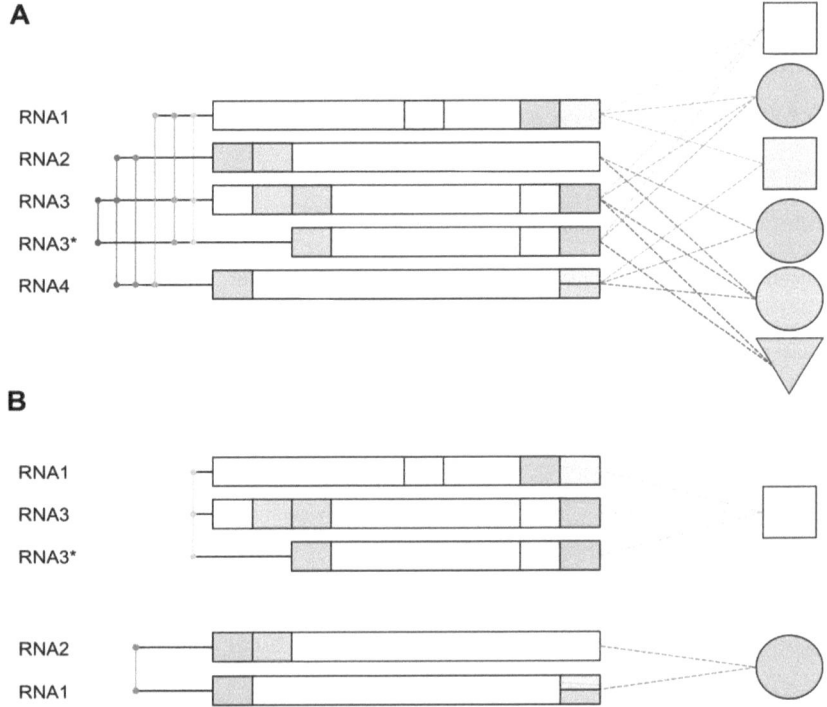

Figure 2.3 Schematic representation of a hypothetical RNA regulon. Five RNA transcripts including different *cis*-regulatory elements are shown together with the corresponding *trans*-acting factors. Different classes of *trans*-acting factors are represented by circles, squares and triangles. (A) An RNA regulon consisting of six different post-transcriptional operons (indicated by different colors on the left). (B) Two exemplary post-transcriptional operons and the regulating *trans*-acting factors are indicated.

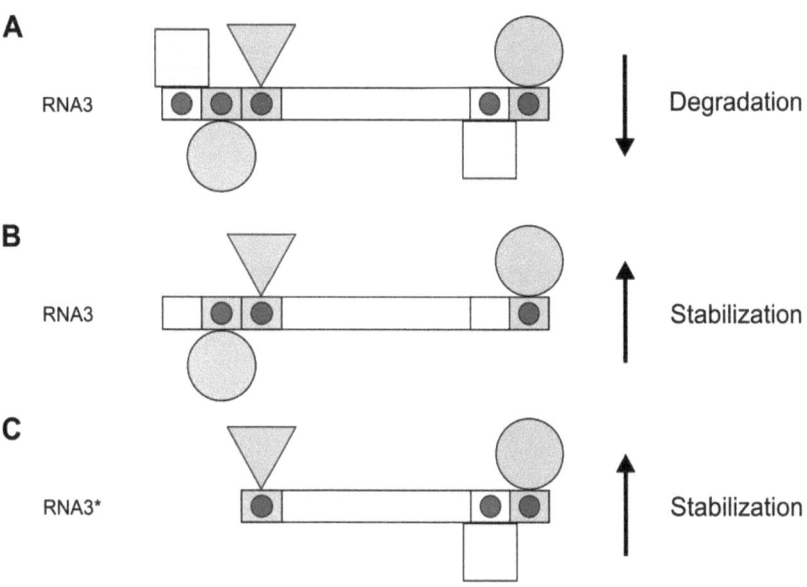

Figure 2.4 Combinatorial control of RNA transcripts. Hypothetical RNA transcripts including different *cis*-regulatory elements are shown together with the corresponding *trans*-acting factors. Different classes of *trans*-acting factors are represented by circles, squares and triangles. (A) In the presence of all *trans*-acting factors the combined regulatory cues lead to degradation of the targeted RNA. (B) In the absence of one of the destabilizing factors (yellow square), e.g. due to differential expression, stabilizing cues dominate. (C) In the absence of destabilizing *cis*-regulatory elements (one yellow and purple box), e.g. through alternative splicing, stabilizing cues dominate. Note that all regulatory cues were assumed to be of equal strength, differing only in their directions.

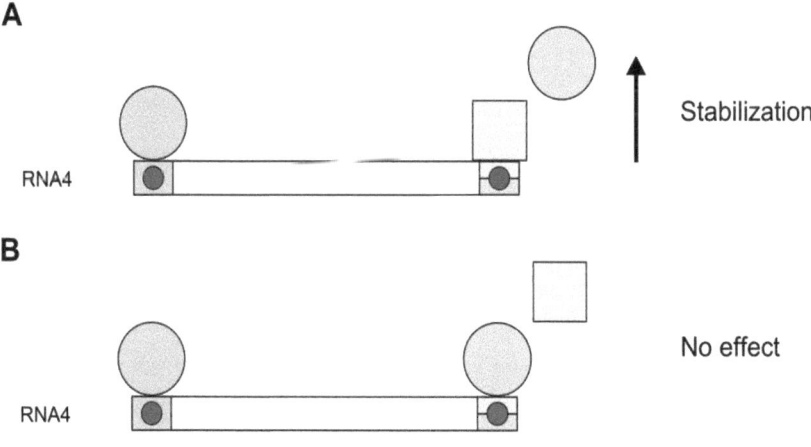

Figure 2.5 Cooperative control of an RNA transcript. A hypothetical RNA transcript including different *cis*-regulatory elements is shown together with the corresponding *trans*-acting factors. Different classes of *trans*-acting factors are represented by circles and squares. Note that two of the *cis*-regulatory elements overlap (light blue and red), so that the respective *trans*-acting factors, one stabilizing and one destabilizing factor, compete for binding. (A) When the stabilizing factor outcompetes the destabilizing factor for binding, the combined regulatory cue is stabilizing. (B) When the destabilizing factor outcompetes the stabilizing factor for binding, the stabilizing cue of the distal *trans*-acting factor is compensated, so that the net effect on the RNA is zero. Note that that all regulatory cues were assumed to be of equal strength, differing only in their directions.

2.2 Post-transcriptional gene regulatory networks

Note: Parts of section 2.2 are abridged from Kanitz, A., and Gerber, A. P. (2010). Circuitry of mRNA regulation. Wiley Interdiscip Rev Syst Biol Med 2, 245–251.

Strikingly, many features of RBP- and miRNA-mediated gene regulation closely resemble those of transcription factors (TFs): While TFs generally bind DNA motifs upstream of a given gene (reviewed in Barrera and Ren, 2006), RBPs and miRNAs typically bind sequence or structural features of mRNA molecules, primarily located in their untranslated regions (Hogan *et al.*, 2008; Hafner *et al.*, 2010b). Moreover, much like RBPs and miRNAs assemble into highly dynamic transient RNP complexes, transcription factors are organized into transcription initiation complexes or "enhanceosomes". Finally, TFs, RBPs and miRNAs often bind targets that code for functionally or cytotopically related proteins (Chu *et al.*, 1998; Lee *et al.*, 2002; Harbison *et al.*, 2004; Gerber *et al.*, 2004; Hogan *et al.*, 2008; reviewed in Halbeisen *et al.*, 2008; Morris *et al.*, 2010). The development and application of genome-wide analysis tools like DNA microarrays revealed fundamental insights into the logic of gene regulatory programs. Chromatin immunoprecipitation (ChIP-Chip) assays have been implemented to systematically map the binding sites of DNA-associated proteins, leading to the identification of transcriptional network motifs (Ren *et al.*, 2000; Iyer *et al.*, 2001; Lieb *et al.*, 2001). For instance, Rick Young and colleagues systematically analyzed the transcription factor binding sites for almost all known transcriptional regulators (203 proteins) in *Saccharomyces cerevisiae* (Lee *et al.*, 2002; Harbison *et al.*, 2004). Likewise, ribonomics approaches (see 2.3) have revealed the (m)RNA targets for dozens of RBPs and thus shed light on the organization of PTGR systems (see 2.3.1). Similarly, several recent studies globally identified targets of individual human miRNAs, either using quantitative proteomics (Baek *et al.*, 2008; Selbach *et al.*, 2008) or RIP-Chip-based approaches, each in the presence or absence of specific miRNAs (see 2.3.1).

In this chapter, we summarize some fundamental motifs of transcriptional and post-transcriptional gene regulatory network (GRNs) circuitries based on selected systematic investigations on the targets of TFs, RBPs and/or miRNAs. Networks, in contrast to linked list and hierarchical data structures, are multidimensional relationships that are neither (exclusively) linear, unidirectional, nor 'rooted'. In the case of GRNs, regulators and targets are interconnected by particular regulatory cues exerted on targets. For the discussed networks, we focus on TFs, RBPs and miRNAs as *trans*-acting factors, although the principles extend to other classes of regulators, such as kinases/phosphatases or methylases/acetyltransferases, as well. A more detailed discourse on general concepts of GRNs can be found elsewhere (reviewed in Milo *et al.*, 2002; Mesarovic *et al.*, 2004; Alon, 2007).

2.2.1 Basic network motifs

The basic network motifs comprise the unidirectional structures: Regulatory cues are exclusively passed down. In biological networks these are mainly constituted by multiple output motifs, i.e. regulatory relationships between one regulator and multiple 'regulatees', as well as multiple input motifs, i.e. the binding of one regulatee by multiple regulators.

2.2.1.1 Multiple output network motifs

As GRNs are non-linear, within each class of regulatory molecules there are instances where one regulator binds to and controls the expression of two or more targets (Figure 2.6). The regulator is usually activated by a signal which could either be an inducer molecule that binds to the regulator or a protein modification of the regulator mediated by a signal-transduction cascade. The frequency of this network motif in GRNs is apparent from elaborate studies analyzing transcription factor binding sites and RBPs in the yeast *Saccharomyces cerevisiae* (Lee *et al.*, 2002; Harbison *et al.*, 2004; Hogan *et al.*, 2008). Lee *et al.* found that each of 106 yeast transcription factors under study bound up to 181 promoter regions ($P < 0.001$), with an average of 38 bound promoter regions per regulator (Lee *et al.*, 2002). Similarly, Hogan *et al.* found that 43 of the 46 RBPs – including two "negative" control proteins – bound more than one RNA target; eleven of them binding less than ten targets, and six of them binding more than a thousand different RNAs, mainly mRNAs (false discovery rate $< 1\%$; Hogan *et al.*, 2008). Although perhaps attributable to the different confidence levels applied for target definition as well as a bias from the RBP selection, it is nevertheless striking that RBPs are associated with an average of about 300 mRNA targets, a number almost ten times as high as the average number of targets for TFs. Whether the larger numbers of RBP targets go along with diminished regulatory impact on individual messages or have other functional implications is not known; yet it illustrates that PTGR networks are at least as densely constructed as TF systems.

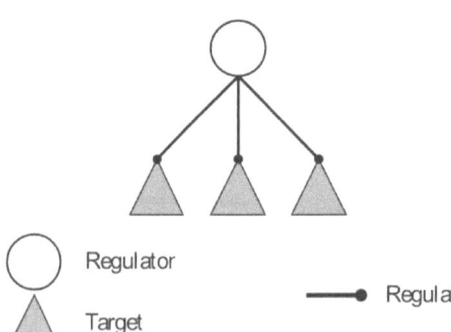

Figure 2.6 Schematic representation of a multiple output network motif. The regulation of three regulatees (i.e. targets) by one regulator is depicted. Adapted from Kanitz and Gerber (2010).

A similar scenario has been observed for miRNAs, where several recent studies have systematically analyzed potential targets for individual human miRNAs, suggesting that each miRNA may bind to and regulate between dozens and thousands of mRNAs (Karginov et al., 2007; Baek et al., 2008; Hendrickson et al., 2008; Selbach et al., 2008; Hafner et al., 2010b). For example, Karginov et al. found that Argonaute 2 (Ago2) proteins associated with 294 unique messages upon overexpression of miR-124 in HEK293 cells (Karginov et al., 2007). Applying the same approach, Hendrickson et al. defined 419 miR-124 target messages, which substantially overlapped with the ones defined by Karginov et al. (Hendrickson et al., 2008). Applying a quantitative proteomics approach, Selbach et al. found the abundance of 1,544 proteins changed in response to miR-124 overexpression (Selbach et al., 2008). Although this number will certainly include secondary effects, it clearly demonstrates the far-reaching consequences of PTGR.

2.2.1.2 Multiple input network motifs

Another hallmark of non-linear GRNs is combinatorial control, which implicates the binding of two or more regulators to a single target (Figure 2.7). Lee et al. mapped almost 4000 individual interactions between transcription factors and promoter regions, providing evidence for the regulation of 2343 of 6270 yeast genes (37%) and an overall connectivity of 1.7 TFs per gene (Lee et al., 2002). Likewise, Hogan et al. mapped 12,000 individual interactions between 46 RBPs and 4,300 mRNAs, indicating a fairly dense overall connectivity (2.8 RBPs per message; Hogan et al., 2008). Extrapolating this to the hundreds of regulatory RBPs present in yeast, each mRNA message might interact with a dozen or more different RBPs on average during its lifetime. These data indicate that the potential for combinatorial controls is considerably higher for RBPs than for TFs, supporting the speculation that PTGR networks are meshed more densely than transcriptional networks.

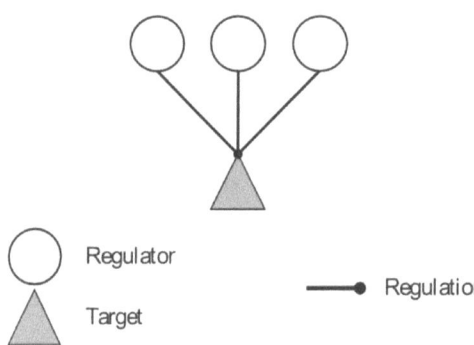

Figure 2.7 **Schematic representation of a multiple input network motif.** The combinatorial regulation of one regulatee (i.e. target) by three regulators is depicted. Adapted from Kanitz and Gerber (2010).

Multiple input network motifs have also been observed for miRNA-mediated PTGR. The expression of VEGFA, for instance, is under the control of at least eight different miRNAs (see 2.4.1). Likewise, synthesis of the tumor suppressors PTEN (Meng *et al.*, 2007; Palomero *et al.*, 2007; Yang *et al.*, 2008) and CDKN1B (see 2.4.2) is controlled by multiple miRNAs.

2.2.2 Autoregulatory, two- and multicomponent loops

The directional regulatory events described above are passed down from one or more regulators to one or more regulatees. A true network, however, is multidimensional, such that each regulator is itself subject to regulation. Consequently, circular motifs are ubiquitously found within GRNs. The simplest form of such a circuit is constituted by an autoregulatory loop, where one regulatory molecule activates (positive feedback loop) or inhibits (negative feedback loop) its own production or activity (Figure 2.8 A). Importantly, autoregulatory loops are thought to be important for the modulation of the response time of gene circuits to a signal and to affect cell-cell variation in protein levels ("noise"). Whereas negative autoregulation generally speeds up the response time of transcriptional gene circuits and reduces cell-cell variation in protein levels, positive autoregulation has slowed response times, leading to enhanced variation (reviewed in Alon, 2007). These circuits may allow rapid adaptation to new environmental conditions (negative loop) or to differentiated states (positive loop).

Global TF binding site analysis in yeast revealed ten high confidence ($P < 0.001$) autoregulatory loops among the interrogated 106 yeast transcription factors (9%; Lee *et al.*, 2002). Likewise, 9 out of the 46 RBPs (20%) surveyed by Hogan *et al.* were reproducibly associated with their own message (Hogan *et al.*, 2008). Interestingly, this number is doubled to 18 autoregulation loops (39% of all studied RBPs) when applying a less stringent cutoff ($FDR < 5\%$). In contrast, miRNAs have not been reported to be organized in autoregulatory loops because miRNAs are not translated but rather target mRNAs in the cytoplasm (reviewed in Djuranovic *et al.*, 2011). Whether the high incidence of such loops in RBP-mediated PTGR has general, systematic implications, possibly beyond the actual pathways they are found in, remains to be analyzed.

More complex regulatory loops are composed of two or more components of a class of regulators (Figure 2.8 B). Lee *et al.* identified three distinct multicomponent loops that consist of two, and in one case of three TFs (Lee *et al.*, 2002). Hogan *et al.* did not explicitly analyze the data from their survey on protein-RNA interactions for the presence of multicomponent loops (Hogan *et al.*, 2008). We therefore re-analyzed the raw data from this survey considering all RNA-protein associations with a false discovery rate of less than 5%. We found that 29 RBPs bound to mRNAs coding for at least one of the 46 RBPs under study. At least 21 of these 29 RBPs are arranged in multicomponent loops. In particular, we identified 13 two-component and 16 three-component loops, involving 16 and 17 different RBPs, respectively (Kanitz and Gerber, unpublished data). This high incidence of multicomponent loops among RBP targets is intriguing and suggests great regulatory potential. One example

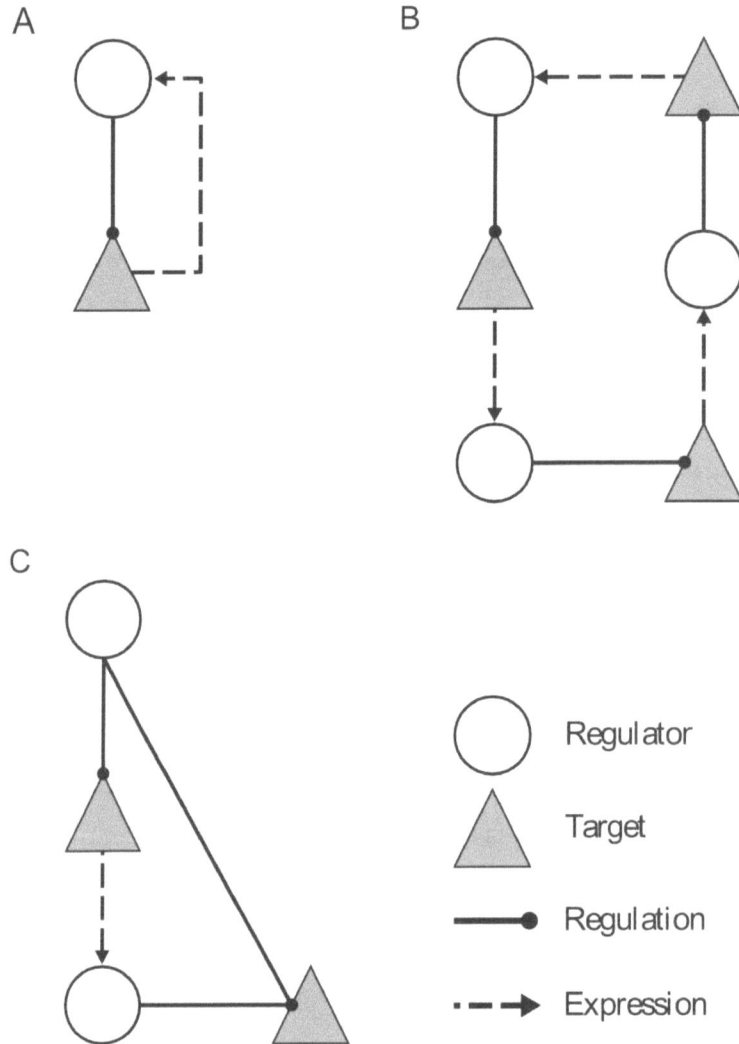

Figure 2.8 Schematic representation of autoregulatory, two- and multicomponent loops. (A) Autoregulatory loop, in which a *trans*-acting factor regulates its own expression. (B) A serial multicomponent loop consisting of three regulators that regulate the expression of each other. (C) A simple feed-forward loop, in which one regulator controls the expression of another regulator, while both regulators control the expression of a common, unrelated target. Adapted from Kanitz and Gerber (2010).

of how multicomponent loops may influence gene regulation is represented by feed-forward loops, a special kind of multicomponent loop which involves the regulation of one or more common targets by two regulators, one of which being under the regulation of the other (Figure 2.8 C). Feed-forward loops can trigger delayed responses to signals, which can be useful to filter out spurious pulses of signals (reviewed in Alon, 2007).

2.2.3 Composite gene regulatory networks

Integration of networks regulated by different classes of *trans*-acting factors leads to an even more complex model of unified 'composite GRNs', which may interact on various levels. Based on selected experimental evidence, some of the concepts of such networks are briefly discussed in this section.

2.2.3.1 RNA-binding proteins versus transcription factors

RBPs are selectively regulated by transcriptional activators or repressors. Among 561 known and predicted yeast RBPs (Hogan *et al.*, 2008), 279 (50%) were targeted by at least one of 106 TFs surveyed by Rick Young and colleagues (Lee *et al.*, 2002). This fraction is considerably larger than the total fraction of regulated genes in the genome (2343 regulated genes out of 6270; 37%). Conversely, the messages encoding TFs require the activity of RBPs for their maturation, decay, localization, and translation. Interestingly, 94 out of the 106 TFs (87%) surveyed by Lee and colleagues were bound by at least one of the 46 RBPs analyzed by Hogan *et al.* (false discovery rate < 5%; Lee *et al.*, 2002; Hogan *et al.*, 2008). Although the study was not exhaustive, this analysis underpins recent observations that RBPs tend to control other gene regulators, such as RBPs and TFs (Figure 2.9 A). Such a "regulator of regulators" concept has recently been established for some human RBPs (Pullmann *et al.*, 2007).

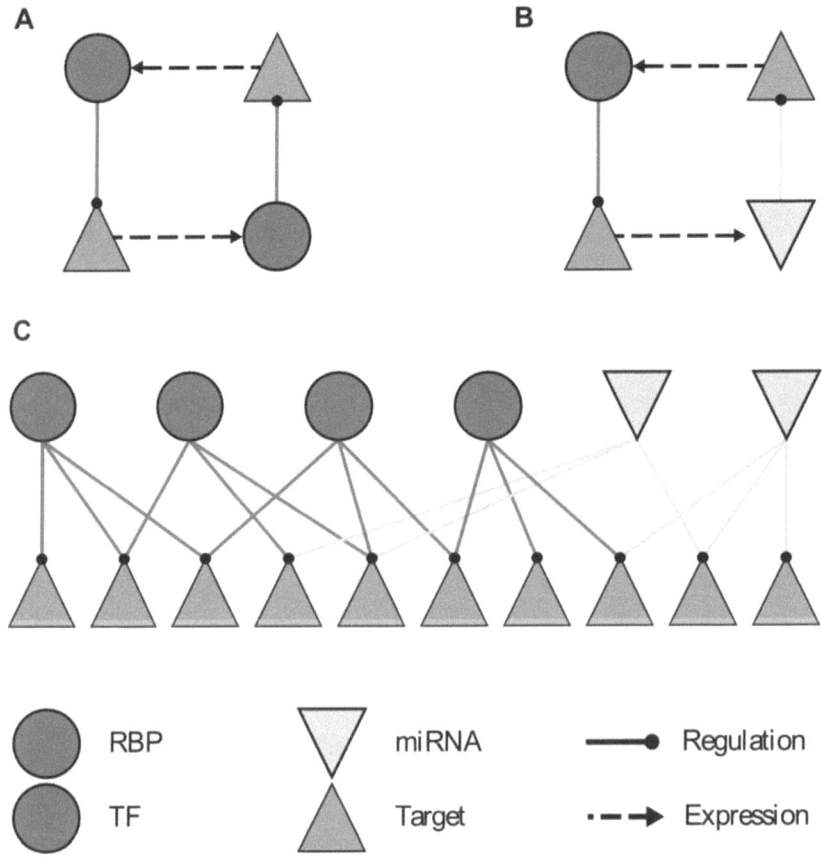

Figure 2.9 Composite gene regulatory network motifs. (A) A two-component loop in which an RNA-binding protein (RBP; blue) regulates the expression of a transcription factor (TF; green), which in turn regulates the expression of the RBP. (B) A two-component loop in which an RBP (blue) regulates the expression of a microRNA (miRNA; yellow), which in turn regulates the expression of the RBP. (C) A unidirectional two-level 'composite regulon' consisting of different classes of gene regulators (TFs, RBPs and miRNAs) which regulate the expression of several target genes in a combinatorial manner. Adapted from Kanitz and Gerber (2010).

2.2.3.2 RNA-binding proteins versus microRNAs

Besides the RNA processing factors required for the biogenesis of the majority of miRNAs, such as the nuclear microprocessor complex and TRBP/Dicer (Denli *et al.*, 2004; Gregory *et al.*, 2004; Chendrimada *et al.*, 2005; Gregory *et al.*, 2005), an increasing body of work suggests that RBPs extensively regulate miRNA expression, either generally or selectively (reviewed in Winter *et al.*, 2009; Krol *et al.*, 2010). In one of the earliest examples of such regulation, Lin28, a cytoplasmic mRNA binding protein, selectively blocks the processing of pre-miRNAs of the let-7 family in human cells (Viswanathan *et al.*, 2008). Interestingly, the message encoding the Lin28 protein is itself regulated by its target miRNA let-7b (Wu and Belasco, 2005), thus providing an example of a RBP/miRNA two-component feedback loop (Figure 2.9 B).

Several studies indicate that RBPs may selectively modulate miRNA function, both synergistically and competitively, to alter translational repression (reviewed in van Kouwenhove *et al.*, 2011). In the first study demonstrating such interplay, ELAV (embryonic lethal, abnormal vision, Drosophila)-like 1 (Hu antigen R; ELAVL1/HuR) relieves the miRNA-mediated repression of CAT-1 mRNA upon stress (Bhattacharyya *et al.*, 2006). For additional examples of crosstalk between miRNAs and RBPs, see 2.4.1 and 2.4.2. Interestingly, bioinformatics analysis of our own ribonomics analyses of the two human Pumilio/Fem-3-binding factor (PUF) family members Pum1 and Pum2 in human cancer cells revealed that conserved Pum recognition elements and miRNA seed sequences were preferentially located in close vicinity among the experimentally identified targets, suggesting extensive crosstalk between the two regulatory systems (Galgano *et al.*, 2008).

The high degree of interplay between transcriptional and PTGR should eventually lead to the characterization of composite GRNs (reviewed in Alon, 2007). For instance, the combination of the multiple-input and multiple-output motifs for functional related gene classes leads to dense and overlapping 'composite regulons' (Figure 2.9 C). Such regulons are widespread phenomena in the control of gene expression at different levels, and can be thought of as gate-arrays, processing multiple inputs from regulators to multiple targets (reviewed in Milo *et al.*, 2002; Alon, 2007). However, in order to understand the functional implications of these composite regulons, not only the connectivity, but also the input function of each regulator (either positive or negative) has to be known, requiring quantitative measurements of the abundance and actions of diverse components of this network. If methodologies can be refined accordingly, such analyses, in the future, will allow a systems-level understanding of the multilayered gene-expression programs.

2.3 Ribonomics methodologies for the systematic identification of basic post-transcriptional gene regulatory network motifs

The advent of global and quantitative analysis tools for the study of gene expression allows the detection and quantification of network motifs in gene regulatory systems. Generally, it appears that principles and structures of transcriptional regulatory networks are also preserved at the post-transcriptional level. However, systems analysis of PTGR is still in its infancy. The development of novel techniques for PTGR network analysis will hence be crucial to obtain sufficient data for the deciphering of the "RNA code". In this chapter, we will summarize the methodologies developed for the unveiling of network motifs, starting with the "top-down" approaches, which allow the identification of the RNA targets of RNA-binding proteins and miRNAs. As these cannot directly identify combinatorial control (i.e. "multiple input") motifs and are thus not in the focus of this work, we will only briefly summarize the available tools, instead focusing on the less well-established "bottom-up" approaches, which allow the identification of proteins and RNAs that bind a specific RNA of interest. Both approaches should allow the analysis of RNA and protein components of purified RNPs by next generation sequencing (reviewed in Wang *et al.*, 2009) and quantitative proteomics methods (Wepf *et al.*, 2009; reviewed in Gstaiger and Aebersold, 2009), respectively.

2.3.1 Top-down approaches: RIP-Chip and related methods

In 1999, the late Robert Cedergren has coined the term 'ribonomics' for the search for RNA genes, their structures and functions (Bourdeau *et al.*, 1999). However, the term now mainly refers to the application of methods aimed at the systematic identification of the RNA components of RBPs. In a pioneering approach, Jack Keene and colleagues have successfully isolated mRNPs by immunopurification of RBPs, followed by the identification of the associated, co-purified mRNA 'targets' using DNA microarrays. Immunopurification relies on the specific immobilization of RNPs on antibody-coupled matrices, and the RBPs are either targeted directly, using RBP-specific antibodies, or via affinity tags that are fused to the RBPs. Targets are defined based on their relative enrichment compared to a suitable control, such as matrices coupled to isotype control antibodies or uncoupled matrices. In analogy to the ChIP-Chip (chromatin immunopurification-microarray) method, which allows the identification of the DNA binding regions of TFs or other DNA-binding proteins (Ren *et al.*, 2000; Iyer *et al.*, 2001; Lieb *et al.*, 2001), the procedure is referred to as RIP-Chip (RBP immunopurification-microarray; Tenenbaum *et al.*, 2000, 2002). The method was a great success and has been employed to determine the RNA targets of more than hundred RNA-binding proteins in mammalian cells, flies, worms, trypanosomes, and particularly in yeast (reviewed in Morris *et al.*, 2010). It also allowed to indirectly determine miRNA targets through the immunopurification of Argonaute proteins, e.g. in human cell lines (Karginov *et al.*, 2007; Hendrickson *et al.*, 2008; Landthaler *et al.*, 2008).

Main limitations of the technique are its inability to discover previously unknown RNAs (as microarrays have to be spotted with specific hybridization probes against potential targets, requiring previous knowledge of their sequences), and the incapacity to directly define the corresponding *cis*-regulatory elements. While the latter may sometimes be circumvented by deducing RBP recognition motifs from sequence comparisons of enriched co-purified messages with the help of pattern recognition/motif discovery software such as MEME (Bailey *et al.*, 2009), this strategy is confined to RBPs with a strong affinity to specific sequence rather than structural or mixed motifs (see the description of PUF proteins in 4.1 for an example of RBPs with high specificity for a particular sequence motif). Furthermore, owing to the dynamics of RNPs and the, consequently, often transient nature of RBP-RNA (or RNA-RNA) interactions, and potential artifacts introduced by the purification procedure, the technique is prone to false negatives and positives.

In a number of recent variations of the technique, the aforementioned obstacles have been addressed, at least in part, by two modifcations: (1) the covalent crosslinking of the bait proteins to the targeted RNAs; and (2) the identification of bound RNAs by high-throughput sequencing methods. Crosslinking procedures rely on either UV light (Greenberg, 1979) or photoactivatable ribonucleoside analogues such as 4-thiouridine (Sontheimer, 1994). The respective techniques, termed HITS-CLIP (high-throughput sequencing crosslinking immunopurification; Licatalosi *et al.*, 2008; reviewed in Darnell, 2010), PAR-CLIP (photoactivatable ribonucleoside enhanced crosslinking and immunopurification; Hafner *et al.*, 2010a, 2010b), and iCLIP (individual nucleotide resolution crosslinking and immunopurification; König *et al.*, 2010, 2011), have been used extensively in the last three years to map in high resolution the RNA targets of several RBPs, including Argonaute proteins (Chi *et al.*, 2009; Hafner *et al.*, 2010b), heterogeneous nuclear ribonucleoprotein (hnRNP) particles (König *et al.*, 2010), T-cell intracellular antigen 1 (TIA1) and TIA1-like 1 (TIAL1; Wang *et al.*, 2010), fragile X mental retardation protein (FMRP; Darnell *et al.*, 2011), and ELAVL1/HuR (Lebedeva *et al.*, 2011; Mukherjee *et al.*, 2011). Moreover, analysis methods (Kishore *et al.*, 2011; Zhang and Darnell, 2011) and databases (Corcoran *et al.*, 2011; Khorshid *et al.*, 2011) have been developed that facilitate the analysis and meta-analysis of such data. The application of these and similar techniques in the coming years will surely contribute immensely to the discovery of multiple output motifs of post-transcriptional GRNs and further our understanding of the RNA code.

2.3.2 Bottom-up: RNA affinity chromatography and related methods

Complementary approaches to RIP-Chip and related methods, focusing on the affinity purification of RNA components of RNPs, should allow both the identification of RNAs bound by *trans*-acting RNAs, as well as *trans*-acting proteins and RNAs binding to messages and non-coding RNAs. Apart from the insights into principles of gene regulatory circuitry, particularly the latter, 'gene-centered' approach might have broad and immediate implications for medical research, as it would allow the convenient and comprehensive identification of post-transcriptional regulators acting on messages of interest, such as oncogenes and tumor suppressors. Although a widely applicable, sensitive, specific and reliable purification strategy for RNAs remains elusive, various promising approaches have been devised and often successfully implemented.

2.3.2.1 Direct RNA affinity chromatography

The most direct approach is represented by RNA affinity chromatography methods that rely on the immobilization of a bait RNA to a matrix and incubation with substrate solution containing proteins or RNAs of interest, such as whole cell extracts, fractions thereof, or purified proteins (reviewed in Kaminiski *et al.*, 1998). The RNA of interest is either *in vitro* transcribed or synthesized, covalently or non-covalently attached to the matrix, and may or may not contain chemical modifications (Grabowski and Sharp, 1986; Bindereif and Green, 1987; Roualt *et al.*, 1989; Caputi *et al.*, 1999; Allerson *et al.*, 2003; Gerber *et al.*, 2004). A method for the immobilization of double-stranded RNA for the isolation of double-stranded

RNA-binding proteins has also been described (Langland *et al.*, 1995). The main disadvantage of such approaches is that RNP formation does not occur *in vitro*, strongly limiting their applicability for the comprehensive characterization of dynamic RNP. Nevertheless, they remain useful for the isolation of individual proteins (Roualt *et al.*, 1989) and relatively static complexes, such as spliceosomes (Grabowski and Sharp, 1986; Bindereif and Green, 1987; Caputi *et al.*, 1999).

2.3.2.2 Purification methods based on antisense hybridization

The specificity of antisense oligonucleotide probes has also been used to characterize the spliceosome: In a variation of the direct RNA affinity chromatography method, immobilized biotinylated antisense RNAs or 2'-O-alkylated RNAs were extensively used to mask specific regions of snRNAs as a means to identify their functional domains and spliceosome architecture (Ruby and Abelson, 1988; Lamond *et al.*, 1989; reviewed in Lamond and Sproat, 1994; Blencowe and Barabino, 1995). A similar approach was later employed to purify *Euplotes aediculatus* telomerase from nuclear extracts via hybridization of immobilized biotinylated 2'-O-methylated (2'-O-Me) RNA complementary to telomerase RNA (Lingner and Cech, 1996), thus extending the use of antisense oligonucleotides for the purification of RNPs beyond the study of spliceosomes. Although the use of immobilized antisense oligonucleotides should allow the purification of native, *in situ*-formed RNPs, the described approaches relied on the incubation of *in vitro*-transcribed bait RNAs with nuclear extracts for RNP assembly, possibly due to the inefficiency of the method. This could possibly be improved by using antisense probes with improved stability, affinity and specificity properties, such as locked nucleic acids (LNAs), morpholinos, or peptide nucleic acids (PNAs). Indeed, in a more recent study, a PNA was coupled to a cell penetrating peptide and a photoactivatable compound and was introduced into cortical neurons where it hybridized with a complementary region in a dendritically localized mRNA. Irradiation with ultraviolet light enabled crosslinking of the PNA to the nearest proteins, and, following RNase digestion, the PNA probe could be captured by an immobilized sense oligonucleotide (antisense to PNA; Zielinski *et al.*, 2006).

2.3.2.3 Aptamer-based purification methods

Aptamers are DNA or RNA sequences that have been selected from random nucleotide libraries for their ability to bind ligands of choice with high affinities (reviewed in Mayer, 2009), usually by systematic evolution of ligands by exponential enrichment (SELEX) (Oliphant *et al.*, 1989; Tuerk and Gold, 1990; Ellington and Szostak, 1990) or related methods. Due to their chemical nature, RNA aptamers can be conveniently used to tag bait RNAs by recombinant DNA technology. Upon expression, RNPs assemble on tagged RNA in an almost native fashion (depending on the site of aptamer insertion), and are amenable to purification via ligand-coated matrices. As such, RNA aptamers constitute perhaps the most promising approach for a 'universal' RNA affinity tag. While the aptamer database (Lee *et al.*, 2004) contains hundreds of different RNA aptamer sequences, so far only a few have been successfully employed for the affinity purification of RNPs: A group of aptamers against the aminoglycoside antibiotic tobramycin has been selected (Wang and Rando, 1995), out of which one has since been established as an RNA affinity tag (Hamasaki *et al.*, 1998) and used to characterize human prespliceosomes (Hartmuth *et al.*, 2002; Hartmuth *et al.*, 2004) and other RNPs (Vazquez-Piazola *et al.*, 2005; Weinlich *et al.*, 2009). An aptamer against streptomycin (Wallace and Schroeder, 1998), another aminoglycoside antibiotic, has also been established as an RNA tag (Bachler *et al.*, 1999; Windbichler and Schroeder 2006; Dangerfield *et al.*, 2006) and applied for the study of the 48S ribosomal subunit (Locker *et al.*, 2006; Locker and Lukavsky, 2007) and group II intron binding proteins (Böck-Taferner and

Wank, 2004). A precursor in yeast ribonuclease P (RNase P) assembly has been identified with the use of an aptamer against Dextran B512 (Srisawat *et al.*, 2001; Srisawat *et al.*, 2002). Finally, an aptamer against streptavidin (Srisawat and Engelke, 2001) has also been used to study RNase complexes (Li and Altman 2002; Xiao *et al.*, 2005; Welting *et al.*, 2007), as well as RNPs and individual proteins associated with small nucleolar RNAs (Lemay *et al.*, 2011), rRNAs (Mohammad *et al.*, 2007), telomerase RNA (Shcherbakova *et al.*, 2009), and AU-rich elements (AREs) (Vasudevan and Steitz, 2007; Vasudevan *et al.*, 2007). One group established a sensitive, gel-free mass spectrometry method for the protein analysis of S1 aptamer-purified RNPs: A tRNA and several mRNA motifs were aptamer-tagged and purified. Analysis of the associated proteins identified proteins that were previously shown to interact with the respective RNAs, as well as several potential new 'interactors' (Butter *et al.*, 2009). Recently, an interesting modification of the S1 aptamer system was reported that involves coupling of the aptamer to a tRNA (Iioka *et al.*, 2011). This report is the first to indicate that the use of scaffolds may enhance both the stability and binding efficiency of aptamers.

2.3.2.4 Indirect protein- and peptide-based purification methods

A more indirect approach exploits protein affinity tags for the purification of RNAs. By fusing bait RNAs to well-established recognition elements of RBPs, it is possible to 'pull down' RNPs via the corresponding, affinity-tagged RBP and a suitable matrix. By co-expression of the RBP, this approach allows the purification of *in situ*-formed RNPs and, furthermore, the direct study of RNP localization when a fluorescent protein is fused to the RBP. The most widely used RBP for these purposes is a coat protein from the R17/MS2 bacteriophage (Bardwell and Wickens, 1990); others include the coat protein of the PP7 phage (Hogg and Collins, 2007a, 2007b), the N antiterminator protein of the λ phage (Czaplinski *et al.*, 2005), the spliceosomal snRNP protein U1A (Brodsky and Silver, 2000;

Takizawa and Vale, 2000) and the eukaryotic initiation factor 4A (Valencia-Burton *et al.*, 2007). As high affinity protein tags are well established and routinely used in many laboratories, and matrices are widely available, this attractive approach has gained some popularity for the purification of RNPs. Affinity tagged MS2 proteins, or fusion proteins thereof (Zhou and Reed, 2003; Beach and Keene, 2008), have been employed, among others, to characterize spliceosomes (Bardwell and Wickens, 1990; Das *et al.*, 2000; Jurica *et al.*, 2002; Jurica and Moore 2002; Zhou *et al.*, 2002; Deckert *et al.*, 2006), mRNA transport/localization proteins or complexes (Gonsalvez *et al.*, 2004), as well as HIV-1 RNA (Kula *et al.*, 2011). Recently, a fusion of the MS2 coat protein to GFP and a streptavidin binding peptide (Keefe *et al.*, 2001) has been described that allows both the tracking (via GFPs) as well as a highly efficient purification of RNPs (via a streptavidin-coated matrix) (Slobodin and Gerst, 2011). While indirect, RBP-mediated approaches are relatively versatile and efficient, a major downside is the indirect nature of the purification method, making it prone to artifacts resulting from the ectopic expression of the protein, cross-reactivity with untagged RNAs, interference with native RNP formation and unspecific binding of RNAs and proteins to the RBP.

2.3.2.5 Bifunctional RNA tag systems

A common obstacle in the widespread use of all of the described approaches is a low signal to noise ratio, mainly due to limited binding specificity and/or low efficiency. Two bifunctional tag systems address these issues by allowing a two-step 'tandem' purification of RNPs: The 'RAT' (RNA Affinity in Tandem) tag contains binding sites for the PP7 coat protein as well as a tobramycin aptamer (Hogg and Collins, 2007a, 2007b), while the 'TRAP' (Tandem RNA Affinity Purification) tag contains an S1 aptamer as well as binding sites for the MS2 coat protein (Nelson *et al.*, 2007). While promising, these tandem tag systems have not gained popularity so far, possibly because their universal use is hindered by the lack of a stabilizing scaffolding structure that could increase the reliability of the systems and ease the selection of integration sites in potential bait RNAs.

2.4 Combinatorial control of cancer-related messages

As diseases often arise as a consequence of dysregulation of one or more genes, it is not surprising that, due to their outlined central roles in the control of eukaryotic gene expression, RBPs and miRNAs are implicated in a large number of human ailments, particularly cancers (reviewed in Calin and Croce, 2006; Lukong *et al.*, 2008; Farazi *et al.*, 2011; van Kouwenhove *et al.*, 2011; Wapinski and Chang, 2011). Here we summarize in detail the extent of combinatorial post-transcriptional control exerted on two cancer-related messages, the angiogenesis factor vascular endothelial growth factor A (VEGFA), and the tumor suppressor cyclin-dependent kinase inhibitor 1B (p27, Kip1)/CDKN1B.

2.4.1 Combinatorial control of the angiogenesis factor vascular endothelial growth factor A

Vascular endothelial growth factor A (VEGFA) is implicated in a wide range of human tumors, as well as several other malignancies, through its function as a potent angiogenic mitogen (see 4.1 for more details). The great variety of mRNA and protein isoforms and the large number of *cis*-binding elements in its mRNA, particularly in its 3'-UTR, render it a prime example of a cancer-related gene that is under extensive combinatorial post-transcriptional control.

The presence of a "tumor-angiogenesis factor" has first been hypothesized in 1971 by Judah Folkman and colleagues based on the description of a mitogen secreted by tumors

(Folkman *et al.*, 1971). In 1983, a factor causing microvascular permeability was partially purified from tumor ascites fluid (Senger *et al.*, 1983). It took another six years, however, until the gene expressing this "vascular endothelial growth factor" was fully purified and eventually cloned (Ferrara and Henzel, 1989; Keck *et al.*, 1989). At the same time, three protein isoforms resulting from alternative splicing were identified, representing the first indication that its expression is regulated post-transcriptionally (Leung *et al.*, 1989). Since then, eight protein isoforms, differing in peptide length and resulting from splicing events in exons 6 and 7, have been characterized (Houck *et al.*, 1991; Poltorak *et al.*, 1997; Lei *et al.*, 1998; Whittle *et al.*, 1999; Lange *et al.*, 2003). Moreover, the use of a distal splice site in exon 8 gives rise to further VEGFA isoforms (Bates *et al.*, 2002) with anti-angiogenic properties, demonstrating the fundamental role of alternative splicing for the function of VEGFA. Changes in the VEGFA co-factor binding domains, resulting from the splicing events, are believed to be the main cause of the differential activity of the various isoforms (reviewed in Harper and Bates, 2008). Additional VEGFA mRNA isoforms may also be generated by the use of an alternative initiation codon (CUG) that is driven by an internal ribosomal entry site (IRES; Huez *et al.*, 2001; Meiron *et al.*, 2001).

The first hints about splicing-independent post-transcriptional control of VEGFA expression came from a 1997 study in which the human VEGFA 3'-UTR was sequenced and five hypoxia-inducible binding sites were identified (Levy *et al.*, 1997). A year later, Levy and colleagues identified the RNA-binding protein ELAVL1/HuR as the first *trans*-acting factor regulating VEGFA expression post-transcriptionally in a hypoxia-dependent manner (Levy *et al.*, 1998) by binding one of the previously identified binding sites (Goldberg-Cohen *et al.*, 2002; Figure 2.10). Similarly, it was shown that the poly(A)-binding protein-interacting protein 2 (PAIP2) stabilizes the VEGFA transcript by binding to its 3'-UTR and interacts with ELAVL1/HuR, thus suggesting a potential cooperative regulation (Onesto *et al.*, 2004).

Recently, ELAVL1/HuR-mediated control of VEGFA expression was shown to contribute to the maintenance of an angiogenic phenotype in tumor-derived endothelial cells, thus underpinning the biological relevance of this interaction (Kurosu *et al.*, 2011).

An early study further reported the presence of a ~125 nt hypoxia-inducible AU-rich stability region in the 3'-UTR, which was shown to be bound by a protein complex of unknown identity (Claffey *et al.*, 1998). One of the components was later identified as the splicing factor hnRNP L, which, under hypoxic conditions, translocates to the cytoplasm and specifically associates with a 21 nt CA-rich element (CARE) within the previously identified AU-rich region (Shih and Claffey, 1999; Figure 2.10). An interesting feedback mechanism of VEGFA expression was revealed by Ray & Fox (2007): While interferon gamma (IFNγ) activates transcription of VEGFA, it also contributes to its translational silencing through the binding of the IFN-gamma-activated inhibitor of translation (GAIT) complex to the VEGFA 3'-UTR. Later it was shown that the VEGFA 3'-UTR region containing the adjacent binding sites for hnRNP L and the GAIT complex represents a conformational switch that allows binding only to one of the two *trans*-acting factors, depending on the status of IFNγ and hypoxia signaling (Ray *et al.*, 2009; Figure 2.10). Recently, Jafarifar *et al.* showed that hypoxia-dependent binding of hnRNP L to the CA-rich element in the VEGFA 3'-UTR led to competitive displacement of miRNAs miR-297, -299, -567 and -605 (Figure 2.10) and consequently derepression of VEGFA expression in tumor-associated macrophages (Jafarifar *et al.*, 2011), providing the first evidence for cooperative regulation of VEGFA expression by a regulatory RBP and miRNAs.

An in-depth analysis of the whole VEGFA transcript revealed the presence of hypoxia-inducible *cis*-regulatory elements also in its 5'-UTR and coding region, and it was demonstrated that their function is dependent on one another (Dibbens *et al.*, 1999).

Subsequently, it was shown that a complex of cold shock domain and polypyrimidine tract binding proteins is able to bind and stabilize both the VEGFA 5'- and 3'-UTRs in a hypoxia-independent manner (Coles *et al.*, 2004). Similarly, the double-stranded RNA-binding protein DRBP76/NF90 was found to bind in the 3'-UTR by Vumbaca and colleagues (2008) and also effected stabilization. Yet another stabilizing factor is constituted by the oncoprotein MDMD2, which like ELAVL1/HuR and hnRNP L translocates to the cytoplasm under hypoxic conditions, where it binds the VEGFA transcript in the AU-rich region (Zhou *et al.*, 2011). In contrast, Ciais and colleagues revealed that the zinc-finger protein TIS11b is able to effect a destabilization of the VEGFA mRNA by binding to a 75 nt region in its 3'-UTR that contains two consensus AU-rich motifs (Ciais *et al.*, 2004).

Apart from the aforementioned study by Jafarifar *et al.* (2011), other miRNAs have also been found to regulate human VEGFA expression post-transcriptionally. In two studies, the group of Zhang You identified several miRNAs that repressed VEGFA expression *in vitro*, although the corresponding recognition motifs in the mRNAs have not been unambiguously identified (Hua *et al.*, 2006; Ye *et al.*, 2008). One of the two VEGFA IRES, driving the expression of the VEGF-121 isoform (Huez *et al.*, 2001), has been shown to be susceptible to regulation by miR-16 (Karaa *et al.*, 2009). Lei *et al.* identified a feedback loop regulating the adaptation of murine tumor cells to different oxygen concentrations in which hypoxia-inducible factor 1 alpha (HIF-1α) suppresses the expression of miR-20b which in turn may regulate both HIF-1α and VEGFA expression (Lei *et al.*, 2009). Two miRNAs, miR-126 and -205, were shown to regulate VEGFA expression and inhibit the growth of lung and breast cancer, respectively, *in vitro* and *in vivo* (Liu *et al.*, 2009; Wu *et al.*, 2009). Two other miRNAs, miR-93 and -200b, were demonstrated to regulate VEGFA expression in diabetes and diabetic retinopathy, respectively (Long *et al.*, 2010; McArthur *et al.*, 2011).

Introduction

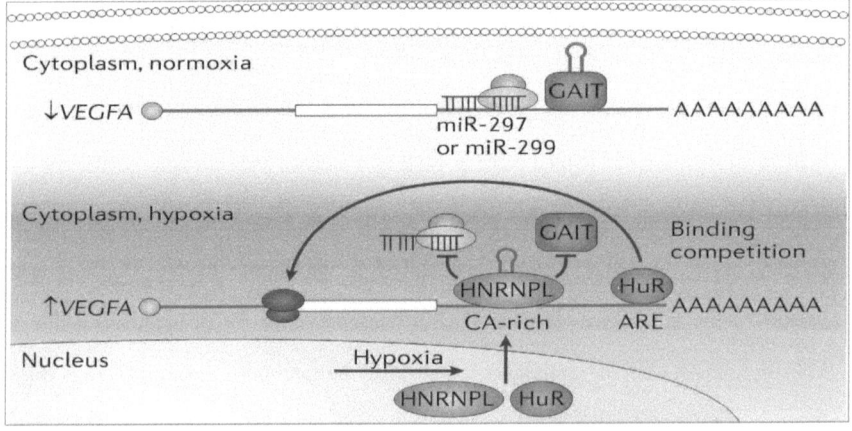

Figure 2.10 Post-transcriptional regulation of VEGFA expression. Some of the identified *cis*-regulatory elements and *trans*-acting factors that regulate VEGFA expression post-transcriptionally are depicted schematically, with a focus on the cooperative regulation that leads to the stabilization of the VEGFA mRNA during hypoxia. See main text for details. Adapted from van Kouwenhove *et al.* (2011).

2.4.2 Combinatorial control of the tumor suppressor CDKN1B/p27

The cyclin dependent kinase (cdk) inhibitor $p27^{Kip1}$/CDKN1B is a tumor suppressor with a key role in the control of the cell cycle. In quiescent cells, the protein is present at high levels where it binds to and inhibits the activity of the cdk4-cyclin D and cdk2-cyclinA/E complexes. Upon mitogenic stimulation in late G1 phase, p27 is targeted for regulated destruction by the Skp1/Cul1/F-box-containing (SCF) ubiquitin ligase complex (Pagano *et al.*, 1995), resulting in the activation of the cyclin complexes and progression of the cell cycle. The protein may be valuable as a prognostic marker or drug target, as its frequent downregulation in tumors is associated with poor prognosis (reviewed in Slingerland and Pagano, 2000; Hershko, 2010). Several post-transcriptional regulators of its expression have been identified that sometimes act in cooperation.

The first indication that CDKN1B expression is under post-transcriptional control came in the mid-nineties: A peak in p27^{Kip1} levels in early G1 phase was not consistent with the stable levels of its mRNA, leading one group to suggest that CDKN1B mRNA may be under translational control (Hengst and Reed, 1996). Later the same year, another group observed that p27^{Kip1} levels drop dramatically upon mitogen-mediated exit from G0 (Agrawal *et al.*, 1996). However, the decrease in protein levels was demonstrated to be independent of its degradation, thus leading them to propose a post-transcriptional regulatory mechanism as well. A role for RNA-binding factors in the regulation of the oscillating p27^{Kip1} levels was finally confirmed in 2000, when it was shown that a U-rich sequence in the CDKN1B 5'-UTR is bound by a number of cycling *trans*-acting factors, specifically the RBPs ELAVL1/HuR, hnRNP C1 and C2, leading to translational activation in proliferating and quiescent cells (Millard *et al.*, 2000). Interestingly, an inhibitory role for the ELAV proteins HuR and HuD in proliferating cells has also been proposed, mediated by the binding to and blocking of an internal ribosome entry site (IRES), which is also situated in the CDKN1B 5'-UTR (Kullmann *et al.*, 2002). Recently, the CUG binding protein was also reported to repress IRES-mediated translation (Zheng and Miskimins, 2011). In contrast, PTB has been proposed to promote cap-independent translation of p27^{Kip1} by interacting with the IRES (Cho *et al.*, 2005). However, other reports call the existence of a functional IRES into question and attribute the observed effects to cryptic promoter activity instead (Liu *et al.*, 2005; Cuesta *et al.*, 2009).

Consistent with the more conventional roles of ELAVL1/HuR, it was recently shown that the protein may also stabilize the CDKN1B message, by binding to several U-rich regions in its 5'- and 3'-UTRs (Ziegeler *et al.*, 2010). In the same study it was also established that multiple mRNA isoforms of CDKN1B are produced by the use of alternative transcription start sites and polyadenylation signals, and that a CU repeat region in the resulting extended 5'-UTR is bound by 41 kDa (kilodalton) stabilizing factor of unknown identity. Another study identified a novel endonuclease that is able to degrade the CDKN1B message by binding to one of its U-rich regions, thus revealing a possible mechanism for the stabilizing role of ELAVL1/HuR, involving the protection of U-rich elements from degradation (Zhao *et al.*, 2000).

While studying oligodendrocyte differentiation in rats, the RBP quaking/QKI was identified as another *trans*-acting factor capable of stabilizing the CDKN1B transcript (Larocque *et al.*, 2005). Due to this role, QKI was recently implicated in the suppression of colon cancer (Yang *et al.*, 2010). Furthermore, the protein was identified to partake in a negative feedback loop involving its direct upregulation by E2F1, followed by the QKI-mediated stabilization of p27, which in turn leads to stabilization of the retinoblastoma protein and suppression of E2F1 (Yang *et al.*, 2011).

By employing an elegant miRNA library screening, Reuven Agami and colleagues identified the first miRNAs suppressing CDKN1B expression, miR-221 and miR-222, and established their roles as potent oncogenes that are upregulated in several cancer cell lines as well as glioblastomas (le Sage *et al.*, 2007). This finding was since corroborated and extended to prostate cancer cell lines (Galardi *et al.*, 2007), myelomas (Felicetti *et al.*, 2008) and hepatocellular carcinomas (Fornari *et al.*, 2008; Pineau *et al.*, 2010), among others. Another miRNA, miR-181a was demonstrated to suppress CDKN1B expression during myeloid cell differentiation (Cuesta *et al.*, 2009). Recently, a third miRNA, miR-148a, was found to regulate CDKN1B and promote proliferation in gastric cancer cell lines, although its levels were frequently downregulated and inversely correlated with CDKN1B levels in gastric cancer tissue samples, indicating a potential antagonistic roles on cancer progression in these particular tumors (Guo *et al.*, 2011).

Two more studies from the Agami lab prominently demonstrated that miR-221/222-mediated suppression of CDKN1B expression is under cooperate control by RBPs: In a 2007 study, it was shown that the RBP Dnd1 is able to bind to two uridine-rich regions located between the two miR-221/222 recognition elements (also see Figure 5.7), thus blocking them from access by the miRNAs in a competitive manner (Kedde *et al.*, 2007). In a more recent study, mitogen-activated phosphorylation of the PUF family RBP Pum1 led to its binding of a *cis*-regulatory element in the proximity of one of the miR-221/222 recognition elements, thus inducing a structural switch that positively influenced the accessibility of the recognition element (Kedde *et al.*, 2010). These findings explain why $p27^{Kip1}$ is able to accumulate in quiescent cells, despite high levels of miR-221/222, and presents an elegant synergistic regulatory mechanism between two different classes of *trans*-acting factors (Figure 2.11).

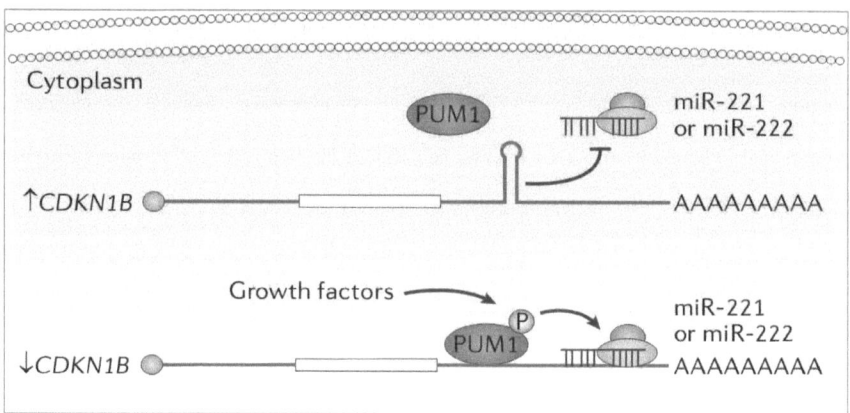

Figure 2.11 Synergistic post-transcriptional control of CDKN1B expression. In the absence of growth factors, Pum1 is in an inactive state and miRNA access to its recognition element is limited. Upon stimulation with growth factors, Pum1 is phosphorylated and binds a recognition element in the CDKN1B 3'-UTR, thus inducing a structural switch that increases the accessibility of the miRNA recognition element. See main text for additional information. Adapted from van Kouwenhove *et al.* (2011).

3   Aims and Outline of the Thesis

The discovery of combinatorial control motifs is a major challenge in the study of post-transcriptional gene regulatory networks. In this work we address this problem by (a) the combination of experimental evidence and bioinformatics predictions to identify two classes of *cis*-regulatory elements in a message of interest (see chapter 4), and (b) the development of a widely applicable method for the identification of *trans*-acting factors that associate with an RNA of interest (see chapter 5), using human cell lines as a model system.

Regarding the first approach, we built upon previous observations which suggest extensive interaction between two classes of post-transcriptional gene regulators, the Pumilio/Fem-3-binding (PUF) family of RNA-binding proteins and the miRNA (miRNA) machinery. To examine the relevance of this finding, we chose to study the 3'-untranslated region of the angiogenesis factor VEGFA, which contains three canonical Pum consensus motifs and is bound by the RNA-binding domains of the two human PUF family members Pum1 and Pum2 *in vitro*. We used five miRNA target prediction services to identify a high confidence miRNA recognition element in the vicinity of the Pum consensus motifs. By using *in vitro* reporter assays, we then assessed the regulatory potential of these *cis*-regulatory elements, as well as the individual and combinatorial regulatory impact of the corresponding *trans*-acting factors.

For the second approach, we set out to develop a method for the enrichment and subsequent compositional analysis of ribonucleoprotein (RNP) particles. By using rational design principles, we aimed to develop a nucleic acid-based RNA tandem affinity tag system that can be fused to RNAs of interest and expressed in human cell lines without impeding native RNP formation. The tagged complexes should then be amenable to purification via a gentle, efficient and highly specific two-step process that is compatible with downstream analytical methods. By using highly sensitive transcriptomics and proteomics approaches, it should then be possible to characterize, in an unbiased manner, their RNA and protein components.

4 Identification of New Post-Transcriptional Regulators of Vascular Endothelial Growth Factor A Expression

4.1 Introduction

Angiogenesis is a multi-step process leading to the formation of vascular structures derived from preexisting blood vessels, either through remodeling or the formation ("sprouting") of new vessels. It involves the induction of microvascular hyperpermeability, breakdown of the vascular basement membrane, recruitment and proliferation of endothelial cells (EC), and the formation of mature blood vessels. Positive and negative regulators of angiogenesis have to be tightly regulated to maintain physiological tissue homeostasis and function. For example, in healthy adult skin, angiogenesis is generally quiescent as angiogenic stimuli are overruled by inhibitory signals. However, environmental insults may tip this balance, leading to initiation of angiogenesis in order to counteract tissue damage. Similarly, excessive angiogenesis in the skin, resulting from dysregulation of one or more of its regulators, is associated with a plethora of pathological conditions, such as psoriasis and other inflammatory dermatoses, autoimmune blistering diseases, and many cancers, most prominently melanoma, basal cell carcinoma and squamous cell carcinoma. Anti-angiogenic therapy therefore holds promise for the treatment of a wide spectrum of human ailments (Detmar, 2000; Carmeliet, 2005).

Cutaneous squamous cell carcinoma (SCC) is the second most common skin cancer in the general population (Lohmann and Solomon, 2001). In contrast to basal cell carcinoma – the most common skin cancer – it is characterized by the risk for metastasis. Incidence of SCC is 60- to 100-times higher among immunosuppressed patients, which makes it the most common cancer following organ transplantation. Invasive SCC develops from atypical keratinocytes, clinically visible as actinic keratosis or Bowen's disease, both considered intraepithelial non-invasive forms of SCC (Hofbauer *et al.*, 2010). Approximately 1% of these intraepithelial lesions develop into an invasive SCC (Schwartz *et al.*, 2008). However, such tumor development requires intense interactions with stromal cells and profound extracellular remodelling. Angiogenesis is an essential part of the malignant phenotype as most tumors are apparently not able to exceed 1-2 mm of diameter without developing new blood vessels (Folkman, 1990). Therefore they produce angiogenic factors at an early point of development.

Vascular endothelial growth factor A (VEGFA) is a homodimeric heparin-binding glycoprotein that mainly acts as a paracrine mitogen, growth and survival factor for ECs, but it also causes vascular permeability, vasodilatation, and various changes in immune cell properties upon binding to its main receptors VEGF receptor-1 and -2. VEGFA has been identified as the predominant tumor angiogenesis factor in the majority of human cancers, including those of the breast, colon, lung and prostate (Ferrara *et al.*, 2003; Hoeben *et al.*, 2004). Invasive SCC also expresses increased levels of VEGFA, particularly in the leading front of the tumor, which is an intuitive site for the induction of angiogenesis (Bowden *et al.*, 2002). VEGFA expression in SCC has not been studied intensively. However in some other cancers, such as head and neck squamous cell carcinoma, increased expression of VEGF has been associated with progression to a more aggressive phenotype, both clinically and in experimental systems (Sauter *et al.*, 1999). Similarly, increased VEGFA expression correlates with greater metastatic potential of melanoma, and its expression is high in melanoma metastases themselves (Salven *et al.*, 1997; Tóth-Jakatics *et al.*, 2000). In the skin, VEGFA is mainly secreted by epidermal keratinocytes. Its expression is upregulated in response to hypoxia (Detmar *et al.*, 1997), activation of epidermal growth factor (EGF) receptor signalling via EGF or transforming growth factor (TGF)-α, and to a number of cytokines including TGF-β, fibroblast growth factor-7 and others (Detmar *et al.*, 1994; Frank *et al.*, 1995). Interestingly, it was shown that heterozygous deletion of the VEGFA 3'-untranslated region (3'-UTR) in mice leads to a two- to three-fold increase in VEGFA levels and embryonic lethality following cardiac failure, thus suggesting the presence of important regulatory elements in its downstream untranslated region (Miquerol *et al.*, 2000). Indeed, VEGFA expression appears to be excessively regulated at the post-transcriptional level, both by RNA-binding proteins and miRNAs (see 2.1.3.2).

Pumilio/Fem-3-binding factor (PUF) proteins belong to a family of regulatory RNA-binding proteins that is conserved among all eukaryotes (reviewed in Spassov and Jurecic, 2002; Wickens *et al.*, 2002; Spassov and Jurecic, 2003). PUF proteins play important roles in a large number of processes, including differentiation and stem cell maintenance, as well as embryonic, germ cell and neural development. For instance, the PUF protein Pumilio, together with the RBPs Nanos and Brat, mediates proper segmentation of fly embryos by inhibiting the translation of *hunchback* mRNA (Wreden *et al.*, 1997; Sonoda and Wharton, 1999). In worms, the PUF protein PUF9 is required for the miRNA let-7-mediated repression of *hbl-1* transcripts, thus regulating the differentiation of epidermal stem cells during larval-to-adult transition (Nolde *et al.*, 2007). The murine PUF protein Pum2 has been shown to localize to neuronal dendrites (Vessey *et al.*, 2006), where it regulates their morphogenesis, as well as synaptic function, at least partly by repressing the translation of eIF4E (Vessey *et al.*, 2010).

The defining characteristic of PUF family members is the presence of a C-terminal RNA-binding domain, referred to as the Pumilio homology domain (Pum-HD). In canonical PUF proteins, it consists of eight imperfect tandem repeats, each of which makes contact with a specific nucleotide in its target *cis*-regulatory element (Wang *et al.*, 2001, 2002), the conserved Pumilio (or Pum) recognition element (PRE). Many PUF proteins further contain a prion-like region that is rich in asparagine and glutamine and may explain their tendency to aggregate and re-localize to stress granules upon their formation (Vessey *et al.*, 2006; Morris *et al.*, 2008; Salazar *et al.*, 2010; Vessey *et al.*, 2010). Due to the unusually high sequence specificity of the Pum-HD, it has been used as a scaffold for engineering RNA-binding domains with engineered sequence specificity (Cheong and Hall, 2006; Dong *et al.*, 2011; Filipovska *et al.*, 2011; reviewed in Lu *et al.*, 2009; Filipovska and Rackham, 2011). Furthermore, the conserved binding motif has been useful as an internal control for ribonomics approaches (see 2.3.1), which is – next to their important biological functions – probably another reason why systematic target analyses have been reported for PUF proteins in humans (Galgano *et al.*, 2008; Hafner *et al.*, 2010b), flies (Gerber *et al.*, 2006), worms (Kershner and Kimble, 2010), trypanosomes (Archer *et al.*, 2009), and yeast (Gerber et al., 2004). The aforementioned characteristics in general and these studies in particular established the PUF family as a prototypical example of regulatory RNA-binding proteins, and they had a considerable impact on the proposition of the RNA regulon theory (see 2.1.4), as it was shown that PUF proteins often bind and presumably regulate functionally and cytotopically related clusters of targets.

In many respects, PUF-mediated post-transcriptional gene regulation (PTGR) strongly resembles the regulation exerted by miRNAs: The conserved AU-rich consensus motif resembles in length and sequence composition the seed sequences of miRNAs (see 2.1.3.2); both PUF proteins and miRNAs preferentially bind in the 3'-UTRs of regulated targets, with a positional bias towards their distal (Pum motifs; Piqué *et al.*, 2008; Kanitz & Gerber, unpublished data), or proximal and distal ends (miRNAs; Gaidatzis *et al.*, 2007); they generally have a repressive effect on target gene expression, with few exceptions (reviewed in Djuranovic *et al.*, 2011; Quenault *et al.*, 2011; Vasudevan, 2011); finally, *Caenorhabditis elegans* contains an unusually high number of different Argonaute (27; Yigit *et al.*, 2006) and PUF proteins (12; Zhang *et al.*, 2011).

There are four human PUF family members, although only two, Pum1 and Pum2, contain the typical Pum-HD containing eight tandem repeats, whereas the other two, KIAA0020/Puf-A and C14orf21 only contain six repeats. Not much is known about the latter two, but Puf-A has previously been identified as a minor histocompatibility antigen (Brickner *et al.*, 2001), and it has recently been shown to re-localize from nucleoli to the nucleoplasm under genotoxic stress, where it then interacts with and modulates the cleavage of PARP-1, a protein involved in DNA damage repair (Chang *et al.*, 2011). No reports exist on C14orf21 function.

The canonical human PUF proteins, Pum1 and Pum2, differ mainly in the presence of an N-terminal stretch of 128 amino acids that is present in Pum1 (molecular weight = 127 kilodalton; kDa), but not Pum2 (molecular weight = 114 kDa; Spassov and Jurecic, 2002), and otherwise share a high degree of similarity, particularly between their homology domains (91% identity, 97% similarity; Spassov and Jurecic, 2002). The consensus sequence for the canonical murine and human PUF proteins is UGUAnAUA (in which n is any nucleotide), as determined by SELEX (White *et al.*, 2001), and ribonomics analyses (Galgano *et al.*, 2008; Hafner *et al.*, 2010b) respectively. Considering the absence of reports demonstrating functional differences between the two proteins and the high overlap between their mRNA targets (Galgano *et al.*, 2008), it is possible that they may act redundantly. Pum1 and Pum2 are ubiquitously expressed, and their levels may oscillate during the cell cycle (Kedde *et al.*, 2010; Kanitz & Gerber, unpublished data). Expression levels of the proteins in cancers were not extensively studied, but Pum1 levels appear to be stable in breast cancers (Szabo *et al.*, 2004; Lyng *et al.*, 2008), but unstable in ovarian cancers (Li *et al.*, 2009). Activity of the Pum proteins may be further modulated by alternative splicing and post-translational modifications, as several isoforms and phosphorylation sites have been found for both proteins. The latter is supported by the recent finding that the Pum1-mediated regulation of CDKN1B/p27 is phosphorylation-dependent (Kedde *et al.*, 2010). However, an in-depth analysis of the role of post-transcriptional and post-translational mechanisms on the regulation of the Pum proteins, or PUF proteins in general, is currently not available.

Here we set out to identify an example of a message that is subject to combinatorial or cooperative control by microRNAs (miRNAs; 2.1.3.2) and the human Pumilio/Fem-3-binding factor (PUF) proteins Pum1 and Pum2, based on our observations that predicted miRNA recognition elements (MREs) are often found in the immediate vicinity of Pum recognition elements (PREs) (Galgano *et al.*, 2008). We chose to focus our studies on human VEGFA, due to the extensive role of post-transcriptional mechanisms in the regulation of its expression and the presence of multiple putative recognition elements for both classes of regulators in its 3'-UTR. We were able to confirm a repressive effect for both Pum proteins and a promising miRNA candidate, microRNA 361-5p (miR-361-5p). While we were not able to study in detail the nature of the potential crosstalk between the regulators, or the significance of VEGFA regulation *in vivo*, we could show that all regulators are present at reduced levels in SCC samples expressing elevated levels of VEGFA.

4.2 Results

4.2.1 The VEGFA 3'-untranslated region contains canonical Pum consensus motifs

The almost 2 kb long sequence of the human VEGFA 3'-UTR (Figure 4.1), the vast majority (>95%) of which is present in all of its known isoforms, contains two regions that are highly conserved among vertebrates, one at its 5'- and the other at its 3'-end. Both regions are lower in GC content compared to the weakly conserved region separating them (GC% approximately 44, 58 and 28 from 5' to 3'), and they contain all unambiguously identified recognition elements for RBPs and miRNAs (Levy *et al.*, 1998; Shih and Claffey, 1999;

Goldberg-Cohen *et al.*, 2002; Hua *et al.*, 2006; Ray and Fox, 2007; Ye *et al.*, 2008; Karaa *et al.*, 2009; Liu *et al.*, 2009; Wu *et al.*, 2009; Ray *et al.*, 2009; Jafarifar *et al.*, 2011; McArthur *et al.*, 2011).

We had previously shown that the RNA-binding domains of Pum1 and Pum2 ('Pum homology domain'; 91% identity and 97% similarity between Pum1 and Pum2) are able to bind the VEGFA 3'-UTR *in vitro*, although RIP-Chip experiments (see 2.3.1) revealed no significant enrichment of the VEGFA message in either Pum1 or Pum2 'pulldowns' of HeLa cell lysates compared to a control (Galgano *et al.*, 2008). The latter was corroborated by a similar study in HeLa cells conducted by Morris *et al.* (2008). However, upon closer inspection of the VEGFA 3'-UTR, we found three canonical Pum consensus motifs (UGUAnATA; Figure 4.1) in its downstream conserved region. The most upstream of these putative PREs, PRE1, was present in the transcript subjected to the *in vitro* binding assays, and could thus explain the observed binding. Recently, a PAR-CLIP-based study (see 2.3.1) in HEK293 cells revealed an association of Pum2 with the downstream conserved region of the VEGFA 3'-UTR at two locations surrounding, but not overlapping, PRE2 and PRE3 (Hafner *et al.*, 2010b). While it appears that the elements PRE1 and PRE2 are mainly conserved among primates (Figure 4.2 A) and humans (Figure 4.2 B), respectively, PRE3 is highly conserved across mammals and marsupials (Figure 4.2 C). Based on these data and observations, we hypothesized that VEGFA might be a target of Pum1 and/or Pum2 and that the lack of a strong association of the VEGFA message in the Pum1/2 RIP-Chip experiments may result from low expression levels of VEGFA in HeLa cells, other *trans*-acting factors inhibiting accesss to the binding sites, or technical limitations.

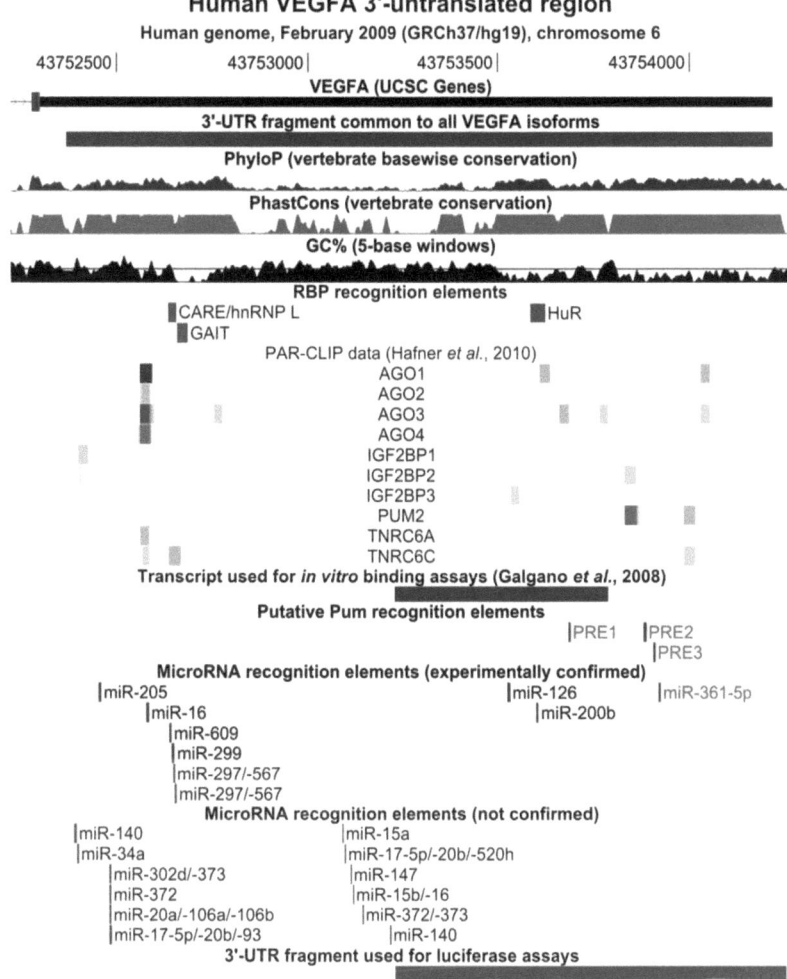

Figure 4.1 Overview of the human VEGFA 3'-UTR. Schematic representation of the genomic locus encoding the 3'-UTR of human VEGFA. The terminal exon of the VEGFA gene, the 3'-UTR fragment common to all isoforms, PhyloP and PhastCons conservation scores, GC content, and the 3'-UTR region subcloned behind a luciferase reporter for use in this study are indicated. Recognition elements of RBPs known to regulate VEGFA expression are depicted. Sequence tag densities of PAR-CLIP experiments for AGO1-4, IGF2BP1-3, PUM2 and TNRC6A/C (Hafner et al., 2010b) are given in gray shades (high density = black, no tags = white). The *in vitro* transcript used in Pum binding assays (Galgano et al., 2008) as well as the putative PREs (red) are indicated. Where available, putative or confirmed seeds and recognition elements for human miRNAs and RBPs regulating VEGFA expression (Levy et al., 1998; Shih and Claffey, 1999; Goldberg-Cohen et al., 2002; Hua et al., 2006; Ray and Fox, 2007; Ye et al., 2008; Karaa et al., 2009; Liu et al., 2009; Wu et al., 2009; Ray et al., 2009; Jafarifar et al., 2011; McArthur et al., 2011) are highlighted, together with the putative miR-361-5p seed (red). The 3'-UTR fragment used for luciferase reporter assays is depicted. Adapted from UCSC genome browser (Fujita et al., 2011).

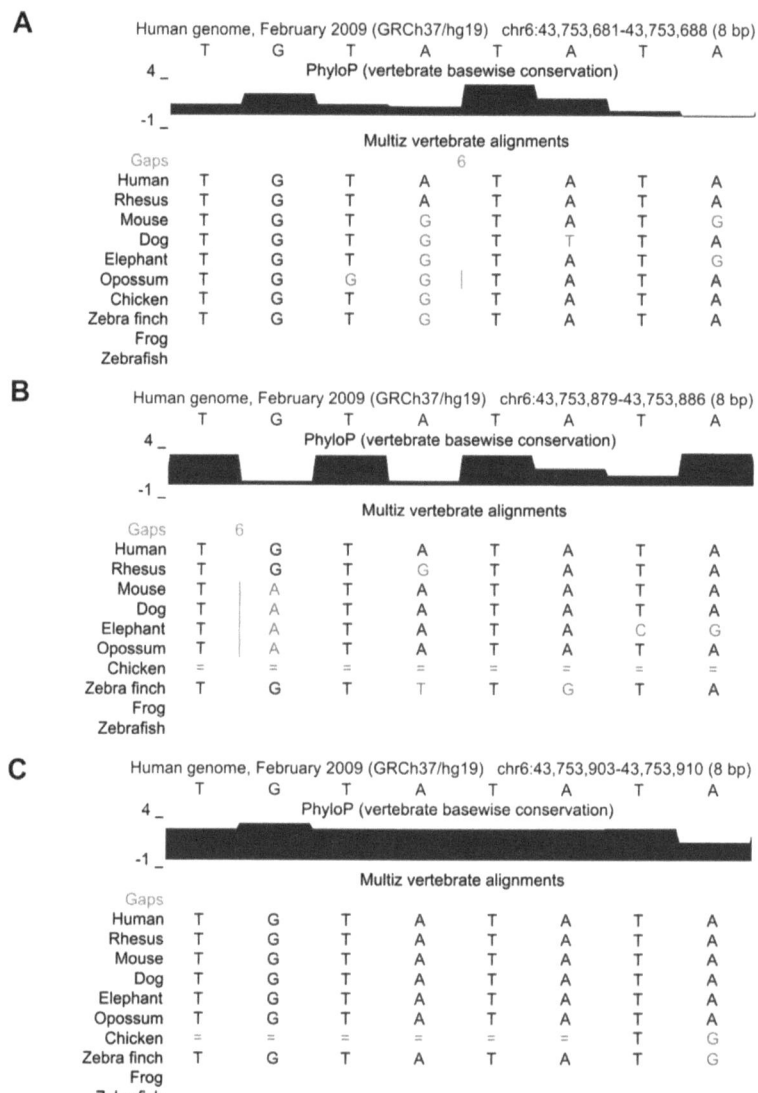

Figure 4.2 Conservation of putative Pum recognition elements. Schematic representation of the three putative PREs in the VEGFA 3'-UTR, PRE1 (A), PRE2 (B), and PRE3 (C). The PhyloP basewise conservation score and sequence alignments for various vertebrates are indicated for each PRE. In the latter, dark blue letters denote nucleotides that are identical with those found in human at the particular positions, while light blue letter indicate those nucleotides that deviate from the human sequence. Insertions in the aligned species are denoted by orange vertical lines; the orange number corresponds to the number of inserted nucleotides in the aligned species with the longest insertion. Light blue double horizontal lines indicate bases that are unalignable. When no alignment is available, spaces are empty. Adapted from UCSC genome browser (Fujita et al., 2011).

4.2.2 VEGFA is a putative target of microRNA 361-5p

Due to the significant enrichment of predicted MREs in close proximity to canonical PREs of experimentally confirmed Pum1/2 targets, we have previously proposed that there may potentially be extensive interplay between Pum- and miRNA-mediated regulation (Galgano *et al.*, 2008). Recently, the finding that Pum1 modulates miR-221/222-mediated regulation of the tumor suppressor CDKN1B/p27 (Kedde *et al.*, 2010) has substantiated this hypothesis. Most miRNA recognition elements (MREs) that have been unambiguously shown to be able to affect human (Figure 4.1) or murine (not shown) VEGFA expression are located in the 5'-conserved region (Jafarifar *et al.*, 2011; Karaa *et al.*, 2009; Lei *et al.*, 2009; Wu *et al.*, 2009; Long *et al.*, 2010); only miR-126 and miR-200b have been demonstrated to bind in the ~730 nucleotide downstream conserved region (Liu *et al.*, 2009; McArthur *et al.*, 2011), albeit relatively far away from the putative PREs (~ 80-380 nt).

In order to find candidate miRNAs that may potentially 'crosstalk' with Pum-mediated regulation of VEGFA expression, we employed five miRNA target prediction services to search for predicted MREs in the VEGFA 3'-UTR: microRNA.org (Betel *et al.*, 2010), TargetScan (Friedman *et al.*, 2009), DIANA-microT (Maragkakis *et al.*, 2009), miRDB (Wang, 2008), and MicroCosm (Griffiths-Jones *et al.*, 2008). The resulting list (see 7.4) contains hundreds of potential binding sites for a large set of miRNAs, covering the entire length of the VEGFA 3'-UTR. Importantly, the density of predicted MREs is considerably higher in the conserved regions of the 3'-UTR (2-3-fold higher compared to the weakly conserved region), particularly in the downstream conserved region (
Table 4.1). This is consistent with the observation that the density of predicted MREs increases towards both ends of a transcript's 3'-UTR (Gaidatzis *et al.*, 2007).

Table 4.1 Density of predicted microRNA recognition elements in the VEGFA 3'-untranslated region. Densities of predicted MREs in the whole 3'-UTR as well as the 3'-UTR regions defined in the main text are given. Target predictions were from the following web services: microRNA.org (MR; Betel et al., 2010), TargetScan (TS; Friedman et al., 2009), DIANA-microT (µT; Maragkakis et al., 2009), miRDB (DB; Wang, 2008), and MicroCosm (MC; Griffiths-Jones et al., 2008). See 7.4 for a full list of miRNAs predicted to target the VEGFA 3'-UTR.

Region	Length (nt)	MREs	Density (MREs/nt)
Conserved region 1	519	150	0,29
Non-conserved region	673	95	0,14
Conserved region 2	733	297	0,41
Whole 3'-UTR (NM_001025366)	1925	545	0,28

Consistent with our bioinformatics analysis (Galgano et al., 2008) and in order to narrow down the list of miRNAs that may co-regulate VEGFA expression together with Pum1/2, we have considered only those predicted MREs whose seed sequences fall within 50 nt of either of the putative PREs, and which have been predicted by at least two algorithms. This analysis revealed 13 potential miRNA:MRE pairs (Table 4.2). Out of these, miR-361-5p emerged as the most likely candidate, as its putative MRE was the only one predicted by all five algorithms. Furthermore, the MRE is situated within 50 nt of not one, but two PREs, including the highly conserved PRE3. Finally, transfection of a miR-361-5p mimic in hypoxia-induced CNE cells has already been shown to reduce VEGFA protein levels, as determined by enzyme-linked immunosorbent assay (ELISA; Ye et al., 2008) – supporting the idea that this miRNA may regulate VEGFA expression. RNAhybrid (Rehmsmeier et al., 2004) calculated a minimum free energy of -22.0 kcal/mol for the interaction between miR-361-5p and the MRE located between nucleotides 1604 and 1625 of the VEGFA 3'-UTR in NM_001025366 (Figure 4.3 A), which is in the range of other miRNAs that have been shown to be able to regulate VEGFA expression *in vitro* (Hua et al., 2006; Ye et al., 2008).

The corresponding *MIR361* gene is encoded on the X chromosome, between exons 9 and 10 of *CHM*/choroideremia (Rab escort protein 1), and gives rise to two mature miRNA species, miRNA-361-3p and the predominant miRNA-361-5p (Figure 4.3 B). The locus is highly conserved among placental mammals, particularly the stem region of the putative precursor miRNA. The mature form, miR-361-5p, has first been isolated from pancreatic islets by Poy and colleagues (Poy *et al.*, 2004), and subsequently from neuroblastoma cell lines (Afanasyeva *et al.*, 2008). No targets for miR-361-5p have been experimentally confirmed so far.

Table 4.2 Predicted microRNA recognition elements in the vicinity of the putative Pum recognition elements. Predicted miRNA seed sequences that fall within 50 nt of either of the three putative Pum recognition elements in the VEGFA 3'-UTR and that are predicted by at least two algorithms are listed. Seed predictions were from the following web services: microRNA.org (MR; Betel *et al.*, 2010), TargetScan (TS; Friedman *et al.*, 2009), DIANA-microT (µT; Maragkakis *et al.*, 2009), miRDB (DB; Wang, 2008), and MicroCosm (MC; Griffiths-Jones *et al.*, 2008). For each miRNA seed, the start and end positions of the seed relative to the start site of the VEGFA 3'-UTR (based on GenBank RefSeq entry NM_001025366.2), the nearby PRE or PREs, the particular algorithms and the total number of algorithms predicting the MRE, as well as the total number of algorithms for which target predictions for the corresponding miRNA were available in the accessed information, are indicated. See 7.4 for a full list of miRNAs predicted to target the VEGFA 3'-UTR.

MicroRNA	Start seed	End seed	PRE	MRE predicted by					Count
				MR	TS	µT	DB	MC	
hsa-miR-548p	1345	1352	PRE1	yes	yes	n/d	yes	n/d	3 / 3
hsa-miR-548d-3p	1376	1383	PRE1	yes	yes		yes		3 / 5
hsa-miR-300	1380	1387	PRE1	yes	yes				2 / 5
hsa-miR-381	1380	1387	PRE1	yes	yes				2 / 5
hsa-miR-590-3p	1388	1395	PRE1	yes	yes				2 / 5
hsa-miR-494	1396	1403	PRE1	yes	yes				2 / 5
hsa-miR-185	1410	1417	PRE1	yes	yes		yes		3 / 5
hsa-miR-300	1568	1575	PRE2/3	yes	yes				2 / 5
hsa-miR-381	1568	1575	PRE2/3	yes	yes				2 / 5
hsa-miR-329	1576	1583	PRE2/3	yes	yes				2 / 5
hsa-miR-362-3p	1576	1583	PRE2/3	yes	yes				2 / 5
hsa-miR-603	1576	1583	PRE2/3	yes	yes				2 / 5
hsa-miR-361-5p	1618	1625	PRE2/3	yes	yes	yes	yes	yes	5 / 5

4.2.3 MicroRNA 361-5p and Pum1/2 may target other angiogenesis-related transcripts

It has been proposed that RBPs and miRNAs often act as master regulators of PTGR (reviewed in Keene, 2007; Mansfield and Keene, 2009; Kanitz and Gerber, 2010). We therefore wondered whether miR-361-5p and Pum1/2 may have additional targets in angiogenesis or related processes. By combining target predictions using the aforementioned five miRNA target prediction services, we have compiled a list of potential miR-361-5p targets (see 7.5 for a partial list). Experimentally verified Pum1 and Pum2 targets were

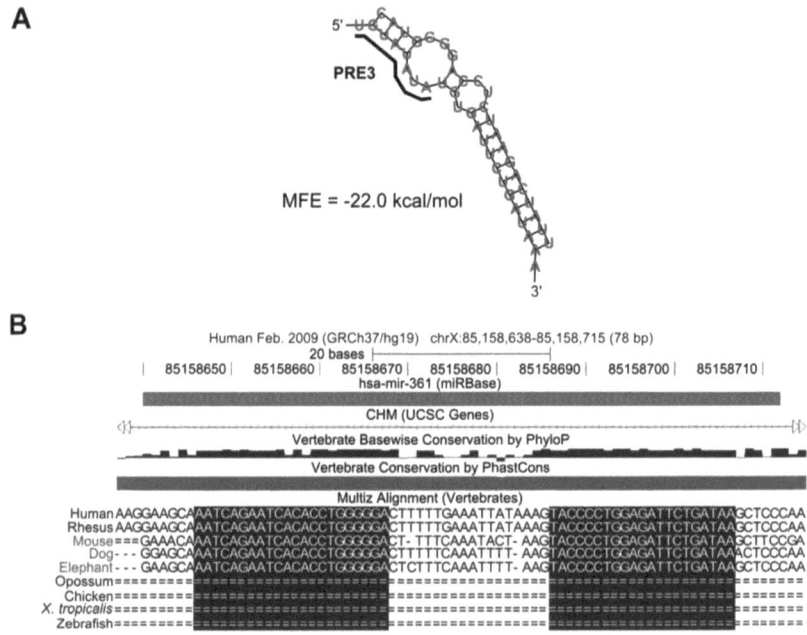

Figure 4.3 MicroRNA 361-5p. (A) Secondary structure of the hybrid between miR-361-5p and the putative MRE within the VEGFA 3'-UTR, as predicted by RNAhybrid (Rehmsmeier et al., 2004). The calculated free energy of the interaction is indicated. Note that the putative PRE3 overlaps with the predicted MRE. (C) The genomic locus encoding hsa-mir-361. The two mature strands, miR-361-5p and -3p are highlighted, and PhyloP and PhastCons conservation scores and alignments with various mammals are indicated. Note that the locus of hsa-mir-361 lies within an intron of the *CHM* gene, encoding Rab escort protein 1. Single horizontal lines indicate deletions relative to the human version, while double horizontal lines indicate bases that are unalignable. Adapted from UCSC genome browser (Fujita et al., 2011).

previously published (Morris et al., 2008; Galgano et al., 2008; Hafner et al., 2010b). Putative targets were then subjected to gene ontology (GO) term annotation using PANTHER (Thomas et al., 2006). Intriguingly, out of the 69 pathways that were significantly ($P < 0.05$) enriched among the putative targets of either of the post-transcriptional regulators, 21 were enriched among all three of them (Table 4.3; see 7.6 for the full list). Strikingly, among these are both angiogenesis ($P = 1.6 \times 10^{-3}$, 1.0×10^{-7}, and 1.2×10^{-8}, for miR-361-5p, Pum1 and Pum2 targets, respectively) and the VEGF pathway ($P = 8.4 \times 10^{-3}$, 1.6×10^{-2}, and 2.8×10^{-3}, for miR-361-5p, Pum1 and Pum2 targets, respectively). A number of other commonly enriched pathways can be linked to angiogenesis or, more specifically, VEGFA function as well, such as EGF, FGF, PI3K, TGF-β and inflammation-related pathways.

Table 4.3 Gene set enrichment analysis of microRNA 361-5p, Pum1 and Pum2 targets. MicroRNA target predictions were from the following web services: microRNA.org (Betel *et al.*, 2010), TargetScan (Friedman *et al.*, 2009), DIANA-microT (Maragkakis *et al.*, 2009), miRDB (Wang, 2008), and MicroCosm (Griffiths-Jones *et al.*, 2008). Experimentally verified Pum1 and Pum2 targets were previously published (Morris *et al.*, 2008; Galgano *et al.*, 2008; Hafner *et al.*, 2010b). Results were pooled and converted to Entrez gene identifiers using the DAVID web service (Huang *et al.*, 2008). Putative targets were compared to a human reference gene list and analyzed for pathway enrichment using PANTHER (Thomas *et al.*, 2006). Pathways commonly and significantly enriched among miR-361-5p, Pum1, and Pum2 targets are listed together with their P values. Angiogenesis and the VEGF pathway are highlighted. See 7.6 for a full list of significantly enriched pathways.

Pathway	miR-361-5p	Pum1	Pum2
PDGF signaling pathway	7.7×10^{-7}	7.0×10^{-7}	1.6×10^{-12}
T cell activation	2.2×10^{-6}	8.9×10^{-6}	6.1×10^{-5}
EGF receptor signaling pathway	7.2×10^{-6}	5.6×10^{-7}	5.7×10^{-10}
p53 pathway	4.3×10^{-5}	1.6×10^{-7}	4.9×10^{-14}
B cell activation	4.9×10^{-5}	3.6×10^{-5}	4.8×10^{-3}
Apoptosis signaling pathway	4.3×10^{-4}	3.6×10^{-4}	1.4×10^{-6}
Ras Pathway	8.2×10^{-4}	1.3×10^{-7}	1.7×10^{-7}
Angiogenesis	**1.6×10^{-3}**	**1.0×10^{-7}**	**1.2×10^{-8}**
Wnt signaling pathway	3.2×10^{-3}	4.3×10^{-4}	3.2×10^{-8}
Inflammation mediated by chemokine and cytokine signaling pathway	4.9×10^{-3}	1.0×10^{-3}	2.1×10^{-2}
Insulin/IGF pathway-MAPKK/MAPK cascade	6.2×10^{-3}	1.9×10^{-3}	3.7×10^{-5}
VEGF signaling pathway	**8.5×10^{-3}**	**1.6×10^{-2}**	**2.8×10^{-3}**
Alzheimer disease-amyloid secretase pathway	9.7×10^{-3}	2.5×10^{-2}	1.3×10^{-3}
p53 pathway feedback loops 2	1.1×10^{-2}	2.2×10^{-2}	1.2×10^{-6}
Oxidative stress response	1.7×10^{-2}	1.0×10^{-3}	5.3×10^{-3}
Integrin signalling pathway	1.9×10^{-2}	5.7×10^{-5}	1.0×10^{-3}
PI3 kinase pathway	2.1×10^{-2}	3.7×10^{-4}	2.2×10^{-7}
Parkinson disease	2.4×10^{-2}	4.3×10^{-4}	4.0×10^{-5}
TGF-beta signaling pathway	2.5×10^{-2}	2.6×10^{-4}	9.0×10^{-7}
FGF signaling pathway	2.6×10^{-5}	9.6×10^{-4}	5.6×10^{-9}
Interferon-gamma signaling pathway	2.9×10^{-2}	5.4×10^{-4}	6.2×10^{-3}

We have also analyzed the manually curated VEGF pathway map at KEGG (Kanehisa *et al.*, 2010) for the list of predicted targets. Among them we found a number of key players, namely PKC, Rac, PI3K, NFAT and cPLA2, which are all predicted to be targets of miR-361-5p by more than one algorithm (Figure 4.4 A), and were further found to be associated with at least one of the Pum proteins (Figure 4.4 B). Even though the prediction algorithms, GO term annotations and transcriptome-wide target identification approaches may be imprecise at the level of individual genes, the strong enrichment of related pathways among the compiled lists suggests that miR-361-5p and Pum1/2 may indeed have additional – and possibly common –

roles in regulating angiogenesis and other VEGF(A)-related functions, both upstream and downstream of the VEGFA/VEGF receptor axis.

4.2.4 Generation and characterization of stable Pum1/2 overexpression cell lines

To study the influence of Pum1 or Pum2 on VEGFA expression, we generated HEK293-derived cell lines stably expressing HA-StrepIII-tandem tagged versions of either the Pum proteins or a control protein (enhanced green fluorescent protein; eGFP) from a single-copy genomic locus upon induction with tetracyclin (Figure 4.5). Immunoblot analysis with an anti-HA antibody confirmed that the resulting cell lines express the introduced coding sequences from the human cytomegalovirus hybrid promoters in a tetracyclin-dependent manner (Figure 4.5 A). Analysis of Flp-In-293-eGFP cells by fluorescence microscopy further indicated that the recombinant eGFP is functional (data not shown). By determining fluorescence levels in cells treated with different tetracyclin concentrations using flow cytometric analysis, we were thus able to determine an 'effective range' of tetracyclin concentrations for the Flp-In-293-derived cell lines: Continuous increases in fluorescence intensities were recorded for tetracyclin concentrations ranging from approximately 0.01 to 1 μg/mL (Figure 4.5 B). As established elsewhere (see 5.2.7), 'leakiness' of the promoter (i.e. basal expression in the absence of inducer) amounts to approximately 10% or less of maximum levels. For Pum1 mRNAs levels in Flp-In-293-Pum1 cell lines treated with different tetracyclin concentrations in the effective range, a similar dose response was observed (Figure 4.5 C). As the used qRT-PCR assay does not differentiate between endogenous and ectopic Pum1 mRNA, it can be concluded that induction with 1.0 μg/mL tetracyclin raises total Pum1 mRNA levels considerably (approximately 4-fold). Finally, we wondered whether recombinant Pum1 co-localizes with endogenous Pum1.

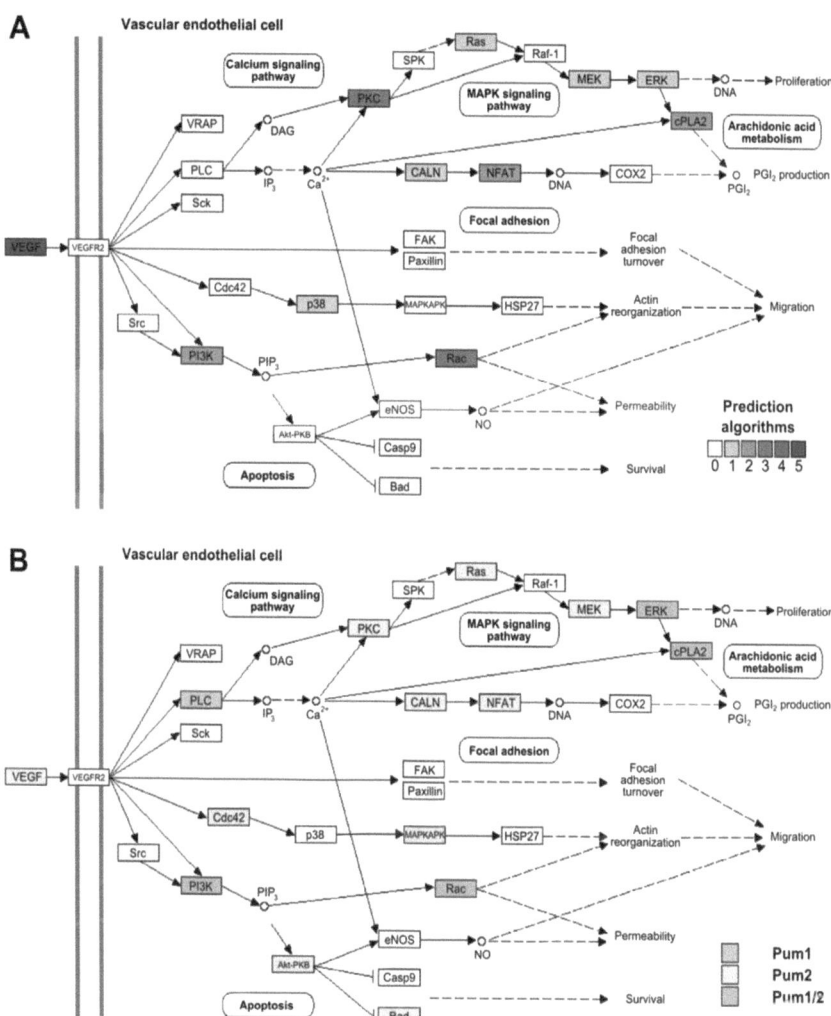

Figure 4.4 Pathway analysis of predicted microRNA 361-5p targets. A manually curated representation of the VEGF signaling pathway available at KEGG (Kanehisa et al., 2010) was color-coded according to (A) the number of algorithms that predict an individual gene to be targeted by miR-361-5p, or (B) whether the corresponding mRNAs were found to be associated with either Pum1 (blue), Pum2 (yellow), or both (green). (A) Target predictions were from the following web services: microRNA.org (Betel et al., 2010), TargetScan (Friedman et al., 2009), DIANA-microT (Maragkakis et al., 2009), miRDB (Wang, 2008), and MicroCosm (Griffiths-Jones et al., 2008). Results were pooled and converted to uniform gene identifiers using the DAVID web service (Huang et al., 2008). (B) Data are from Morris et al. (2008; Pum1 only), Galgano et al. (2008), and Hafner et al. (2010b; Pum2 only).

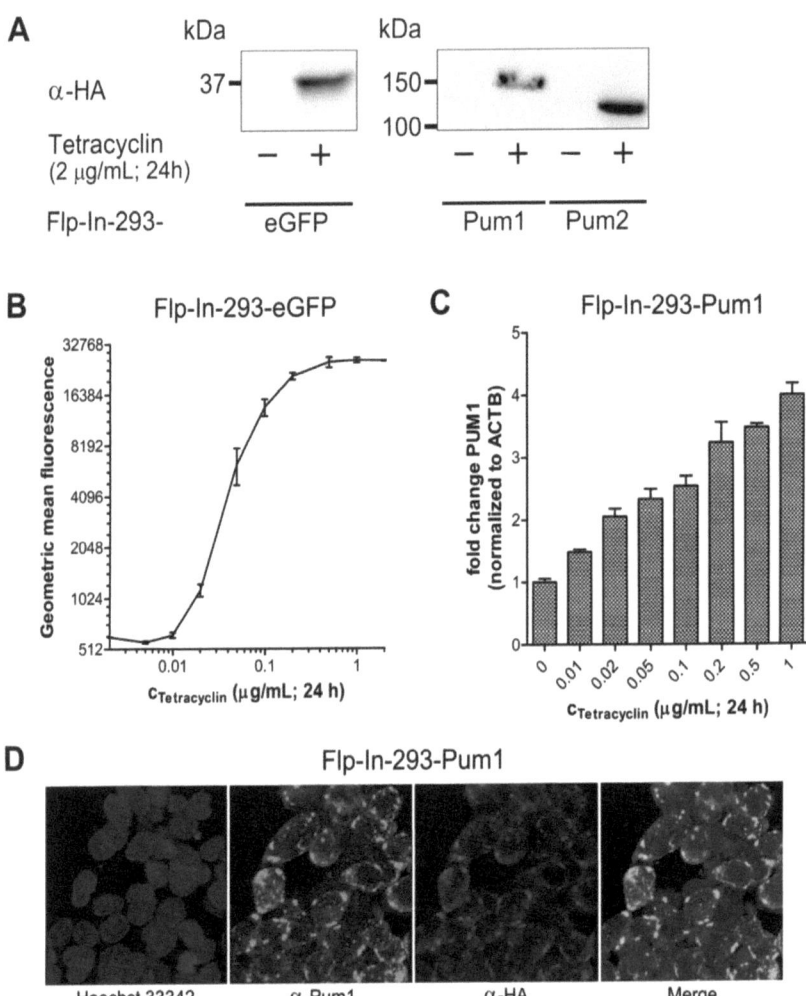

Figure 4.5 Characterization of Flp-In cell lines. (A) Immunoblot analysis of Flp-In-293-eGFP, -Pum1 and Pum2 cell lines. Lysates of tetracyclin- (+) or ethanol-treated (EtOH; -) Flp-In-293-eGFP, -Pum1 or –Pum2 cells were subjected to immunoblot analysis with an anti-HA antibody (clone HA-7). (B) Flow cytometry analysis of Flp-In-293-eGFP cells treated with different concentrations of tetracyclin. Geometric mean fluorescence levels are plotted against the tetracyclin-concentration. (C) Pum1 transcript levels in Flp-In-293-Pum1 cells treated with different concentrations of tetracyclin were assayed by qRT-PCR analysis. Fold changes ± S.D. in expression levels with respect to vehicle-treated (0 µg/mL) cells are plotted. ACTB was used as a reference. Experiments were performed in triplicate. (D) Flp-In-293-Pum1 cells were treated with tetracyclin (1 µg/mL; 24 h) and sodium arsenite (0.5 mM; 45 min). Subsequently, cells were stained with Hoechst 33342 dye (blue) and Pum1 (A300-201A; green) and HA (clone HA-7; red) antibodies and analyzed by fluorescence microscopy. Data in (B and C) are from a single experiment.

Immunocytometric analysis with anti-HA and anti-Pum1 antibodies to detect recombinant and endogenous Pum1 revealed that both proteins are present ubiquitously in the cytoplasm under normal conditions, resulting in diffuse cytoplasmic staining (data not shown), thus making it difficult to assess whether ectopic Pum1 is localized correctly. However, it has been reported that Pum1 localizes to stress granules under oxidative stress conditions, which appear as clearly visible cytoplasmic foci when analyzed by microscopy (Morris *et al.*, 2008). In order to induce stress granule formation, we therefore treated Flp-In-293-Pum1 cells with sodium arsenite in addition to tetracyclin. Immunocytochemistry confirmed that under oxidative stress conditions both proteins co-localize to distinct foci in the cytoplasm (Figure 4.5 D). Taken together, these data indicate that the generated cell lines produce the recombinant HA-StrepIII-tagged proteins upon treatment with inducer in a dose-responsive manner and that, at least in the case of eGFP and Pum1, the proteins are likely folded properly.

4.2.5 Transfection of small RNAs

In order to study the effects of altered miRNA levels in cellular systems, it is paramount to establish efficacious transfection methods for small RNAs. We therefore assessed the ability of HEK293, A431 and HaCaT cells to take up miRNA mimics and antisense inhibitors by transfecting them with increasing amounts of Cy3-labeled Pre- and Anti-miR control constructs, respectively, followed by flow cytometric analysis (Figure 4.6 A to C). The resulting data revealed that Cy3 labeled constructs were incorporated by more than 80% of all cell lines across the whole range of concentrations (Figure 4.6 B) and with fluorescence intensities increasing in a dose-respondent manner (Figure 4.6 C). Differences in the fractions of transfected cells were small between cell lines and constructs (< 20%). By transfecting the used cells lines with Pre- and Anti-miR constructs, we should therefore be able to alter the levels of miRNAs of interest as desired.

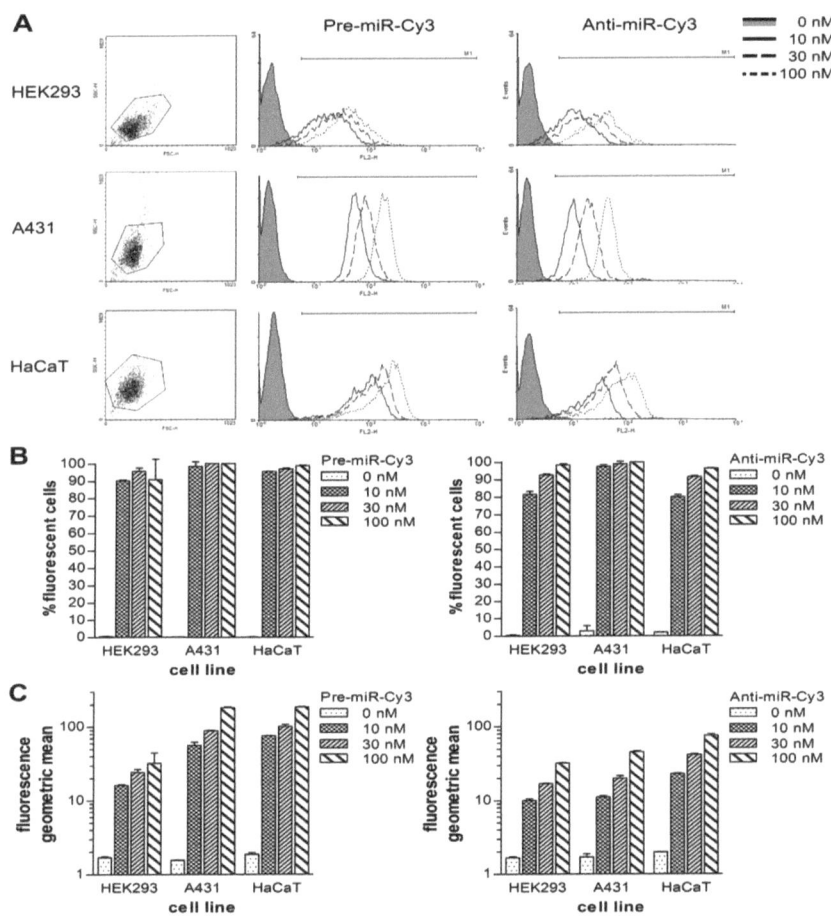

Figure 4.6 MicroRNA mimics and antisense inhibitors are readily taken up by HEK293, A431 and HaCaT cells. Cells were transfected with either 0 (mock), 10, 30 or 100 nM of Cy3-labeled Pre-miR or Anti-miR constructs and analyzed by flow cytometry. Experiments were performed in triplicate. (A) For each cell line, dot plots of mock-transfected cells (left panel) indicate the populations subjected to fluorescence analysis. The fluorescence distributions of gated cells are plotted for a single replicate, both for Pre-miR (middle) and Anti-miR-transfected cells (right). The fluorescence thresholds for positive cells are indicated (M1). The efficiencies of transfection are represented as the mean fractions of fluorescent cells ("M1 positive cells") (B) and the mean geometric means (C) within the gated populations ± S.D.

4.2.6 The putative Pum and microRNA 361-5p recognition elements in the VEGFA 3'-UTR possess regulatory potential

Due to the generally high degree of interconnectivity within post-transcriptional regulatory networks (see 2.2), the extensive post-transcriptional regulation that has already been shown to be exerted on the VEGFA transcript (see 2.4.1), and the high occurrence of predicted MREs in the conserved region surrounding the putative MRE (see 7.4), it is possible that other *trans*-binding factors might affect the binding potential of miR-361-5p, Pum1 and/or Pum2. Thus we reasoned that it may be beneficial to preserve potential RNA recognition elements in our experiments. We therefore cloned the entire downstream conserved region of the VEGFA 3'-UTR behind the coding sequence of *Renilla* luciferase under the control of a Simian virus 40 (SV40) promoter, on a plasmid (psiCHECK-2) further encoding a firefly luciferase for normalization purposes (Figure 4.7 A and B). As a positive control, we also generated luciferase reporter constructs bearing the CDKN1B 3'-UTR, which is known to be cooperatively regulated by Pum1 and miR-221/222 (Kedde *et al.*, 2010). Additionally, we generated variants of both reporters in which either the PREs or MREs were mutated. In order to avoid competition between the reporter and endogenous VEGFA, we performed the assays in Flp-In-293 cells, as the parental human embryonic kidney (HEK293) cell line expresses low levels of VEGFA (Liang *et al.*, 2002).

In order to assess the regulatory potential of the total VEGFA 3'-UTR fragment as well as that of the PRE and MRE motifs of interest, we transfected Flp-In-293 cells with the wild type and mutated VEGFA and CDKN1B reporters or unmodified psiCHECK-2 (Figure 4.7 C). While relative *Renilla* activity was significantly decreased by almost half in cells transfected with the wild type CDKN1B reporter when compared to the psiCHECK-2 control ($P = 2.82 \times 10^{-5}$; unpaired *t*-test, two-tailed), no such change was observed for the wild type VEGFA reporter. These data indicate that, in the tested cell line, the combined post-transcriptional regulation exerted on the VEGFA 3'-UTR fragment is roughly neutral, whereas in the CDKN1B 3'-UTR negative regulators (i.e. repressors) appear to prevail. Mutation of either the PREs or the miR-221/222 MREs in the CDKN1B 3'-UTR leads to an

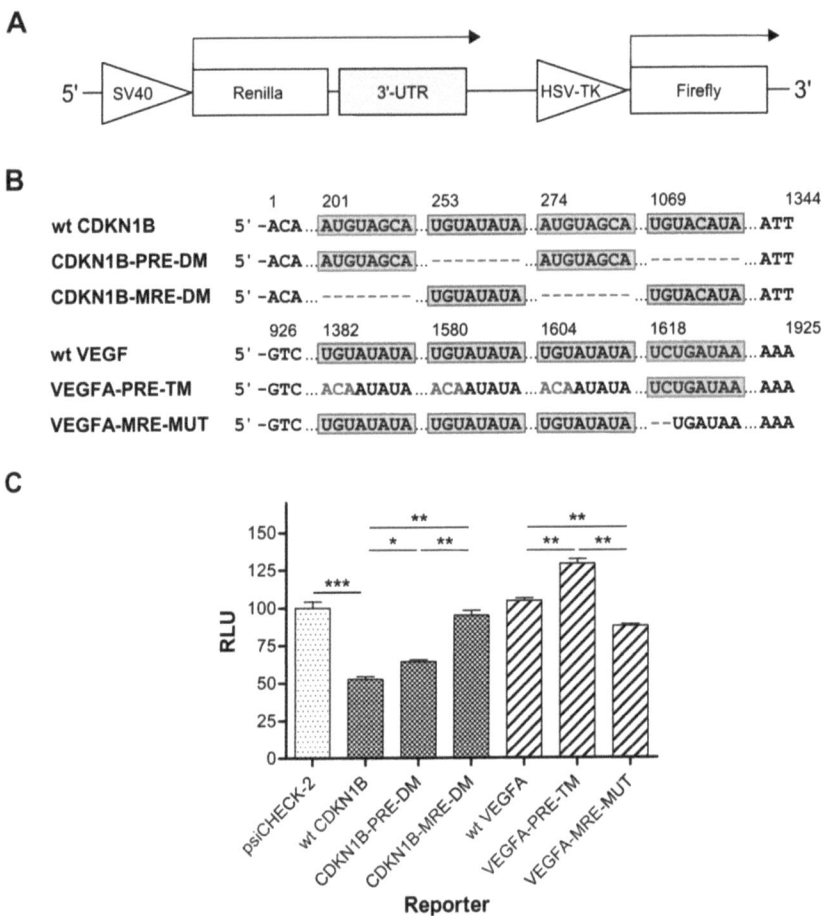

Figure 4.7 **Effect of the CDKN1B and VEGFA 3'-untranslated regions and their putative Pum and microRNA recognition elements on the activity of a luciferase reporter.** (A) Schematic representation of the luciferase reporter constructs (psiCHECK-2 vector, Promega). 3'-UTR fragments (yellow) are fused to *Renilla* luciferase, which is under the control of a simian virus 40 (SV40) promoter. Firefly luciferase, under the control of a herpes simplex thymidine kinase (HSV-TK) promoter, is used for normalization. (B) Overview of the wild type (wt) and mutated CDKN1B and VEGFA 3'-UTR fragments used for the generation of luciferase reporter constructs. miR-221 (blue) and putative Pum (brown) and miR-361-5p (green) recognition elements are highlighted, and the first and last three nucleotides of the fragments are given. Numbers denote the positions of the first residues of each motif relative to the start of the respective 3'-UTRs according to GenBank RefSeq mRNA entries NM_004064. 2 (CDKN1B) and NM_001025366.2 (VEGFA). Note that while the miR-361-5p MRE was mutated at three consecutive nucleotides (see 7.1.2 for the sequences of oligonucleotides used for the mutagenesis), only two mutations are present in the seed sequence. (C) Flp-In-293 cells were transfected with either of the indicated luciferase reporters. Mean ratios of *Renilla* and firefly luciferase activities (relative luciferase units; RLU) were normalized to those of psiCHECK-2-transfected cells. Four independent experiments were performed at least in triplicate. Mean values ± S.E.M. from a representative experiment are plotted. Two-tailed, unpaired *t*-tests were used to calculate P values (one, two and three asterisks denote P values <0.05, <0.01, and <0.001, respectively).

increase in relative luciferase activity (1.23- and 1.81-fold higher than the wild type 3'-UTR, respectively; P = 0.0103 and 0.0011; unpaired t-test, two-tailed) in cells transfected with the respective reporters. While the mutation of the PREs in the VEGFA 3'-UTR also resulted in elevated relative luciferase activity (1.24-fold higher than the wild type 3'-UTR; P = 0.0053; unpaired t-test, two-tailed), the opposite was observed for the mutation of the putative miR-361-5p MRE (approximately 1.19-fold lower than the wild type 3'-UTR; P = 0.0025; unpaired t-test, two-tailed). These data indicate that the studied 'motifs' in the VEGFA 3'-UTR appear to possess intrinsic regulatory potential and thus represent *cis*-regulatory elements.

4.2.7 Pum1, Pum2 and microRNA 361-5p repress the expression of VEGFA 3'-UTR reporters

To study the effects of elevated Pum1 or Pum2 levels on the expression of VEGFA, we transfected Flp-In-293-derived cell lines overexpressing either eGFP, Pum1 or Pum2 with the luciferase reporters and compared relative luciferase activities (Figure 4.8). For all CDKN1B and VEGFA reporters, overexpression of either Pum1 or Pum2 resulted in significant drops in *Renilla* activities of approximately 25% compared to eGFP overexpression (e.g. P = 0.0016 and 0.0033 for the wild type CDKN1B reporter, for Pum1 and Pum2 respectively; e.g. P = 2.9 x 10^{-6} and 0.0253 for the wild type VEGFA reporter, for Pum1 and Pum2 respectively; unpaired t-test, two-tailed). When disregarding the higher base level activities of the PRE mutants compared to wild type reporters (see 4.2.6), differences between luciferase activities in cells transfected with wild type or PRE mutants were very moderate (approximately 6% and 13% increase in the PRE mutants for CDKN1B and VEGFA, respectively; P = 0.2065 and 2.3 x 10^{-4}; unpaired t-test, two-tailed). Mutation of MREs did not result in significant changes compared to the wild type reporters for both genes.

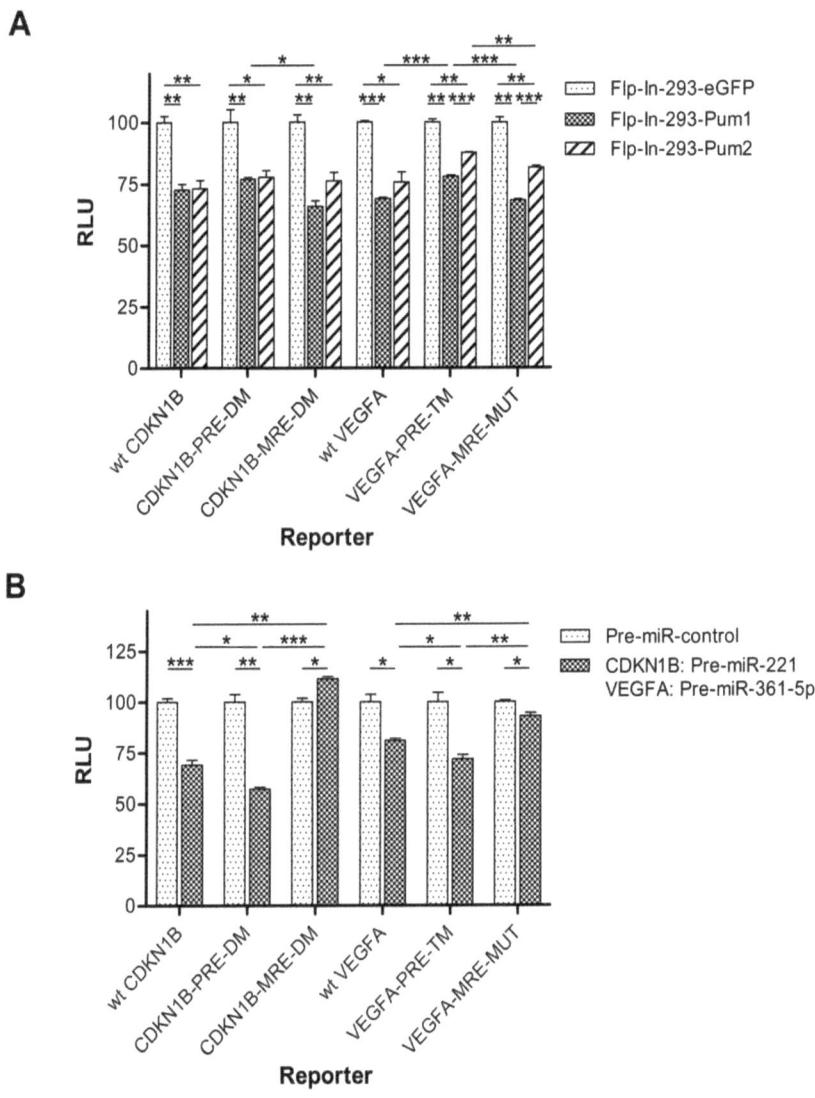

Figure 4.8. Effect of elevated Pum1, Pum2 and microRNA 221 or 361-5p levels on the activity of VEGFA and CDKN1B 3'-UTR luciferase reporters. (A) The indicated Flp-In-293-derived cell lines were transfected with either of the indicated luciferase reporters. (B) Flp-In-293 cells were co-transfected with the indicated luciferase reporters and Pre-miR microRNA mimics. Mean ratios of *Renilla* and firefly luciferase activities (relative luciferase units; RLU) were normalized to those of Flp-In-293-eGFP (A) or Pre-miR-control-transfected cells (B). Two independent experiments were performed in triplicate. Mean values ± S.E.M. from one experiment are plotted. Two-tailed, unpaired *t*-tests were used to calculate P values (one, two and three asterisks denote P values <0.05, <0.01, and <0.001, respectively).

While the effects of overexpressing Pum2 were generally lower (between 1% and 11%) when compared to Pum1 overexpression, differences only reached significance in the VEGFA PRE and MRE mutant reporters (P = 2.2 x 10^{-4} 1.9 x 10^{-4}; unpaired *t*-test, two-tailed). The results indicate that the VEGFA is susceptible to repression by Pum1 and Pum2 which appears to be at least partly mediated by one or more of the putative PREs.

Analogously, the effects of elevated levels of miR-361-5p on VEGFA were studied by co-transfecting Flp-In-293 cells with the VEGFA luciferase reporters and either a miR-361-5p mimic or a control, or, for control purposes, with the CDKN1B reporters and either a miR-221 mimic or control. For cells transfected with the wild type reporters, the addition of miRNA mimics led to significant decreases in luciferase activities compared to the addition of control constructs, both for CDKN1B and VEGFA (31% and 19%, respectively; P = 7.3 x 10^{-4} and P = 0.0270; unpaired *t*-test, two-tailed). Effects were even more pronounced when PREs were mutated (43% and 28%, respectively; P = 0.0057 and P = 0.0130; unpaired *t*-test, two-tailed), whereas mutation of the MREs fully (CDKN1B; 11% increase for the miR-221 mimic) or partly (VEGFA; 8% decrease for the miR-361-5p mimic) abolished this effect.

Taken together, the results show that miRNA-361-5p is able to repress VEGFA reporter expression via the putative MRE in the respective 3'-UTR.

4.2.8 The repressive effects of Pum proteins and microRNA 361-5p on VEGFA 3'-UTR reporter activity are additive

To assess whether miRNA-361-5p and Pum proteins exert combinatorial or even cooperative regulation on the VEGFA 3'-UTR, we co-transfected the Flp-In-293-derived cell lines expressing recombinant eGFP, Pum1 or Pum2 with either miRNA mimics or controls and wild type and mutated VEGFA or CDKN1B 3'-UTR luciferase reporters and compared luciferase activities (Figure 4.9). We found that for both the CDKN1B and VEGFA reporters, most of the data fit with a model in which the effects of simultaneously elevated Pum and miRNA levels add up to the sum of their individual effects (Table 4.4). This suggests that the

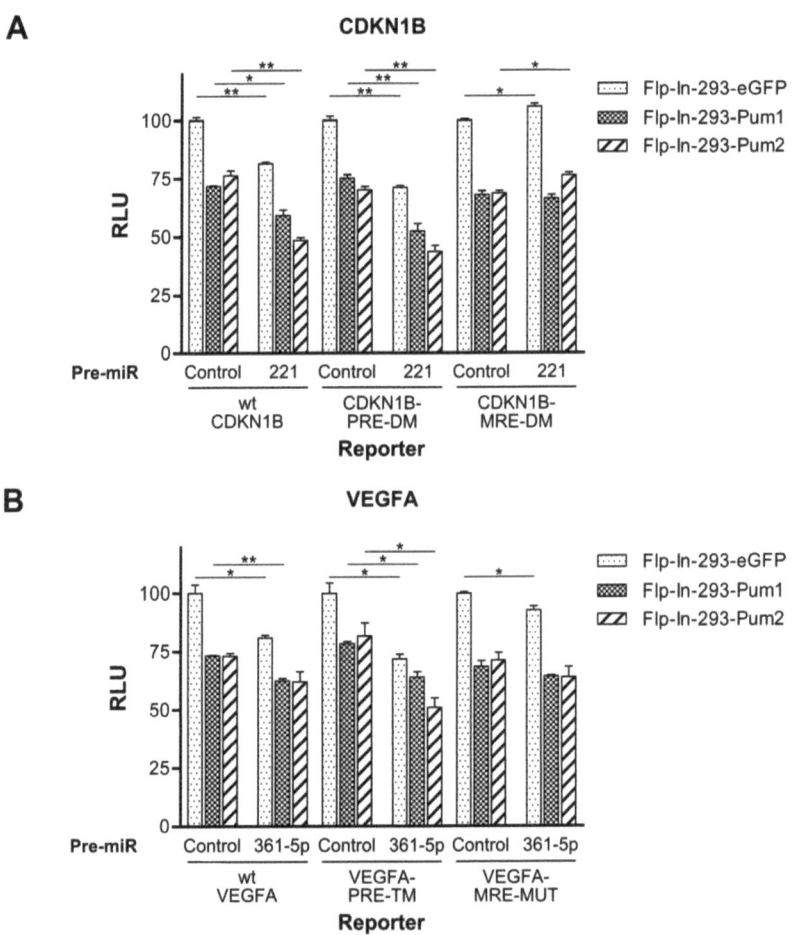

Figure 4.9 Effect of simultaneously elevated levels of microRNA 221 or 361-5p and Pum1 or Pum2 on the activity of VEGFA and CDKN1B 3'-UTR luciferase reporters. The indicated Flp-In-293-derived cell lines were co-transfected with either of the indicated CDKN1B (A) or VEGFA (B) luciferase reporters and Pre-miR microRNA mimics. Mean ratios of *Renilla* and firefly luciferase activities (relative luciferase units; RLU) were normalized to those of Pre-miR-control-transfected Flp-In-293-eGFP cells. Two independent experiments were performed in triplicate. Mean values ± S.E.M. from one experiment are plotted. Two-tailed, unpaired *t*-tests were used to calculate P values (one and two asterisks denote P values <0.05 and <0.01, respectively).

Table 4.4 Observed and expected reductions in reporter activities. The effects of simultaneously elevated miRNA and Pum1 or Pum2 levels (observed; obs.), expressed as percent reduction compared to the baseline (transfection of Flp-In-293-eGFP with Pre-miR-control), are compared to the sum of their individual effects (expected; exp.). Fold changes and P values (unpaired t-test, two-tailed) are indicated. Significance was assumed for $P < 0.05$.

Reporter	% reduction (observed)	% reduction (expected)	Fold change obs vs. exp	P value
miRNA and Pum1				
wt CDKN1B	40.77 ± 4.01	41.42 ± 1.38	0.98	0.8095
CDKN1B-PRE-DM	47.59 ± 5.65	46.23 ± 2.94	1.03	0.7267
CDKN1B-MRE-DM	33.43 ± 2.74	27.83 ± 3.75	1.20	0.1040
wt VEGFA	37.52 ± 1.69	40.70 ± 1.91	0.92	0.0715
VEGFA-PRE-TM	36.05 ± 4.00	43.69 ± 3.86	0.83	0.0660
VEGFA-MRE-MUT	35.70 ± 0.71	36.37 ± 5.18	0.98	0.7885
miRNA and Pum2				
wt CDKN1B	51.48 ± 2.05	37.70 ± 3.85	1.37	0.0060
CDKN1B-PRE-DM	56.22 ± 4.46	49.92 ± 2.94	1.13	0.1220
CDKN1B-MRE-DM	23.43 ± 2.08	27.11 ± 3.12	0.86	0.1396
wt VEGFA	37.97 ± 7.71	40.80 ± 2.87	0.93	0.5961
VEGFA-PRE-TM	49.01 ± 7.05	41.38 ± 10.78	1.18	0.2777
VEGFA-MRE-MUT	36.18 ± 7.80	33.87 ± 6.40	1.07	0.7015

regulation of the 3'-UTRs of CDKN1B and VEGFA by Pum1/2 and miR-221, or miR-361-5p respectively, is largely independent of one another. The only condition in which the data significantly deviates from such an additive model is when Flp-in-293-Pum2 cells were co-transfected with the wild type CDKN1B reporter and a miR-221 mimic. Here, an approximately 51% reduction in reporter activity was observed, which is a considerably stronger repression (1.37-fold; $P = 0.0060$; unpaired t-test, two-tailed) than would have been expected from their individual effects (38% reduction). These data better fit a model that assumes synergism between Pum2 and miR-221. As for the VEGFA 3'-UTR, the observed reductions differed noticeably, yet not significantly, from expected ones only for the reporter in which the PREs were mutated. For this reporter, the effects of a simultaneous increase in miR-361-5p and Pum1 levels are somewhat weaker than would be expected from applying a strictly additive model (0.83-fold; $P = 0.0660$; unpaired t-test, two-tailed). In contrast, the observed reduction in reporter activity upon increasing miR-361-5p and Pum2 levels appears to be stronger than the expected additive effects (1.18-fold; P = 0.2777; unpaired t-test, two-tailed).

Taken together, these data indicate that both the VEGFA and CDKN1B 3'-UTRs are regulated by Pum1, Pum2 and either miR-361-5p (VEGFA) or miR-221 (CDKN1B) in a combinatorial manner. Most of the observed data fit well with an additive model of co-regulation, thus suggesting that the regulatory effects exerted on the 3'-UTRs are largely independent of one another.

4.2.9 Endogenous VEGFA expression is regulated by microRNA 361-5p

To check whether endogenous VEGFA levels could be affected by miR-361-5p, we chose two different cell lines derived from human skin that are known to express high levels of VEGFA: the epidermoid squamous cell carcinoma-derived A431 cell line, and HaCaT cells, keratinocytes derived from normal skin that transformed spontaneously *in vitro*. First, we determined the expression levels for endogenous miR-361-5p and VEGFA mRNA in these cell lines using quantitative reverse transcription PCR (qRT-PCR; Figure 4.10). While miRNA expression did not differ between the two cell lines (fold difference between A431 and HaCaT = 1.04 + 0.27 – 0.21), VEGFA levels were significantly higher in A431 cells compared to HaCaT cells (3.69 + 0.38 – 0.35; P = 2.1×10^{-7}; unpaired t-test, two tailed).

On the protein level, the baseline VEGFA secretion rates of mock-transfected cells were approximately 3262 ± 585 and 1314 ± 152 pg/mL after 24 hours, for A431 and HaCaT cells respectively (ratio ~ 2.5), as determined by ELISA (Figure 4.11).

We then determined VEGFA levels in the culture supernatants of both cell lines

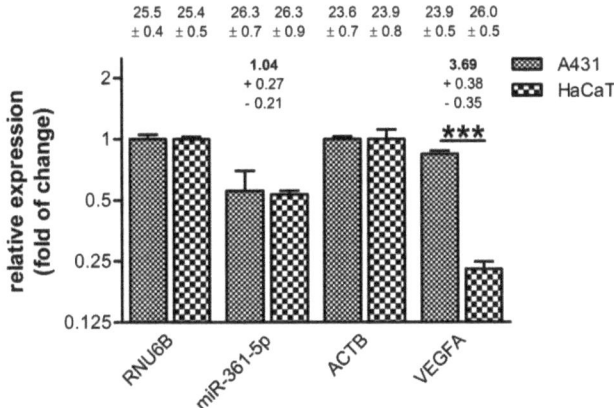

Figure 4.10 VEGFA and miR-361-5p are expressed in A431 and HaCaT cells. qRT-PCR analysis of VEGFA and miR-361-5p expression in A431 and HaCaT cells. Fold differences ± S.D. in expression levels with regards to the references (ACTB and RNU6B, for VEGFA and miRNA-361-5p, respectively) are plotted. Fold differences ± S.D. between A431 and HaCaT expression levels are given for miR-361-5p and VEGFA. Mean C_T values ± S.D. are indicated above each column. Data represent at least three independent experiments performed in triplicate. Two-tailed, unpaired t-tests were used to calculate P values (the triple asterisk denotes a P value <0.001).

transfected with different amounts of miRNA-361-5p mimic or control using ELISA (Figure 4.11 A and B). While we only observed a slight decrease in VEGFA levels in HaCaT cells (up to ~11% when transfecting 30 nM; P = 0.0502; unpaired t-test, two-tailed), VEGFA levels were significantly decreased in A431 cells (up to ~30 % when transfecting 10 nM; P = 0.0063; unpaired t-test, two-tailed) when comparing transfection of miR-361-5p mimic and control.

Conversely, VEGFA levels were not affected in A431 cells transfected with increasing amounts of miRNA-361-5p antisense inhibitor when compared to cells transfected with a control (Figure 4.11), while in HaCaT cells we detected elevated VEGFA levels for all tested antisense inhibitor concentrations (up to ~39% when transfecting 10 nM; P = 0.0150; unpaired t-test, two-tailed). Taken together, these findings demonstrate that altered levels of miR-361-5p may affect the rates at which VEGFA is secreted, suggesting that the miRNA is able to repress endogenous VEGFA expression *in vitro*.

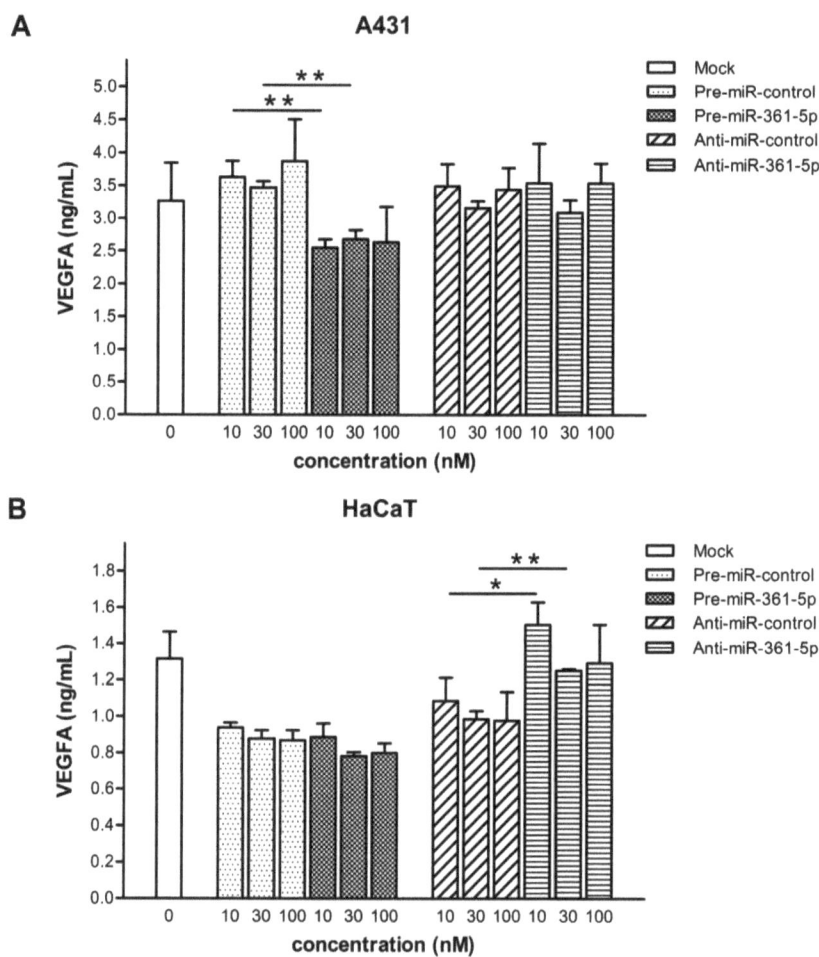

Figure 4.11 Impact of altered miRNA-361-5p levels on secretion of endogenous VEGFA. A431 (A) and HaCaT (B) cells were transfected with the indicated concentrations of miR-361-5p mimic, antisense inhibitor, or controls. Culture supernatants were analyzed for VEGFA protein content using a human VEGFA ELISA. Three independent experiments were performed in triplicate. Mean values ± S.D. from a representative experiment are plotted. Two-tailed, unpaired *t*-tests were used to calculate P values (one and two asterisks denote P values <0.05 and <0.001, respectively).

4.2.10 MicroRNA 361-5p is down-regulated in cutaneous squamous cell carcinoma

Having established that miR-361-5p is expressed and able to regulate VEGFA expression in skin-derived cell lines, we hypothesized that its expression may potentially be downregulated in SCC and that it may thus contribute to the initiation or maintenance of high VEGFA expression. We therefore measured the expression of miR-361-5p and several other VEGFA-regulating miRNAs in five samples of SCC obtained from patients and in five healthy skin samples using qRT-PCR.

First, we investigated whether VEGFA expression was indeed increased in the SCC samples by assessing VEGFA mRNA levels with two different assays, one for exon 3 and the other one for the downstream conserved region in the 3'-UTR. As expected, we found that VEGFA mRNA levels were around two-fold higher in the SCC samples compared to healthy control samples (fold difference between SCC and healthy skin for the exon 3 assay: 2.27 + 2.61 – 1.22; $P = 0.0472$; unpaired t-test with Welch's correction, two-tailed; $P = 0.0556$; Mann-Whitney $U = 3$, $n1 = n2 = 5$, two-tailed; fold difference between SCC and healthy skin for the 3'-terminal assay: 2.12 + 4.50 – 1.44; $P = 0.1846$; unpaired t-test with Welch's correction, two-tailed; $P = 0.2222$; Mann-Whitney $U = 6$, $n1 = n2 = 5$, two-tailed; Figure 4.12 A). Moreover, data correlated very well for both assays ($r = 0.83$, $P = 0.0015$; Spearman's rank correlation, one-tailed; Table 4.5). Interestingly, we found that the VEGFA 3'-terminus was expressed at significantly lower levels than the coding region (fold difference between VEGFA exon 3 and VEGFA 3'-terminus assays of 2.72 + 1.34 – 0.90; $P = 0.0098$; unpaired t-test with Welch's correction, two-tailed; $P = 0.0115$; Mann-Whitney $U = 17$, $n1 = n2 = 10$, two-tailed; Figure 4.12 B and C).

We then measured the expression of Pum1, Pum2, miR-361-5p, its 'host gene' CHM and the known VEGFA-regulating miRNAs miR-20b, miR-34a, miR-93, miR-126 and miR-205. In healthy skin samples, the average expression levels of miR-20b (~62-fold down) and miR-205 (~51-fold up) strongly deviated from that of the reference RNA (RNU6B), while for all other miRNAs differences stayed within an order of magnitude (Figure 4.12 B). Of note, miR-361-5p levels (~3.6-fold lower than RNU6B) were very consistent between samples and correlated well with CHM mRNA levels (r = 0.53, P = 0.0587; Spearman's rank correlation, one-tailed; data not shown). The protein coding genes were all present at levels comparable to those of VEGFA, with Pum2 being expressed at ~2.2-fold higher levels than Pum1. In the SCC samples, Pum1, Pum2, CHM and miR-361-5p levels were significantly decreased compared to healthy skin samples (fold difference between SCC and healthy skin for the PUM1 assay: 0.58 + 0.18 − 0.14; P = 0.0175; unpaired t-test with Welch's correction, two-tailed; P = 0.0079; Mann-Whitney U = 0, n1 = n2 = 5, two-tailed; fold difference between SCC and healthy skin for the PUM2 assay: 0.36 + 0.08 − 0.07; P = 0.0001; unpaired t-test with Welch's correction, two-tailed; P = 0.0079; Mann-Whitney U = 0, n1 = n2 = 5, two-tailed; fold difference between SCC and healthy skin for the CHM assay: 0.40 + 0.53 − 0.23; P = 0.0456; unpaired t-test with Welch's correction, two-tailed; P = 0.0952; Mann-Whitney U = 4, n1 = n2 = 5, two-tailed; fold difference between SCC and healthy skin for the miR-361-5p assay: 0.44 + 0.33 − 0.19; P = 0.0220; unpaired t-test with Welch's correction, two-tailed; P = 0.0159; Mann-Whitney U = 1, n1 = n2 = 5, two-tailed), whereas the other tested miRNAs either did not exhibit considerably reduced expression levels or, as seen for miR-126, even appeared to be upregulated (fold difference between SCC and healthy skin of 3.72 + 10.99 − 2.78; P = 0.0699; unpaired t-test with Welch's correction, two-tailed; P = 0.0952; Mann-Whitney U = 4, n1 = n2 = 5, two-tailed; Figure 4.12 A and C).

Importantly, miR-361-5p levels exhibited the strongest inverse correlation with VEGFA levels across all samples (r = -0.58 and -0.60, P = 0.0408 and 0.0333; Spearman's rank correlation, one-tailed; for the exon 3 and 3'-terminal assays, respectively; Table 4.5). No other miRNA or protein coding gene passed an r ≤ -0.5 threshold. Out of the Pum proteins, Pum1 levels correlated considerably stronger with VEGFA levels (r = -0.45 and -0.41, P = 0.0935 and 0.1221; Spearman's rank correlation, one-tailed; for the exon 3 and 3'-terminal

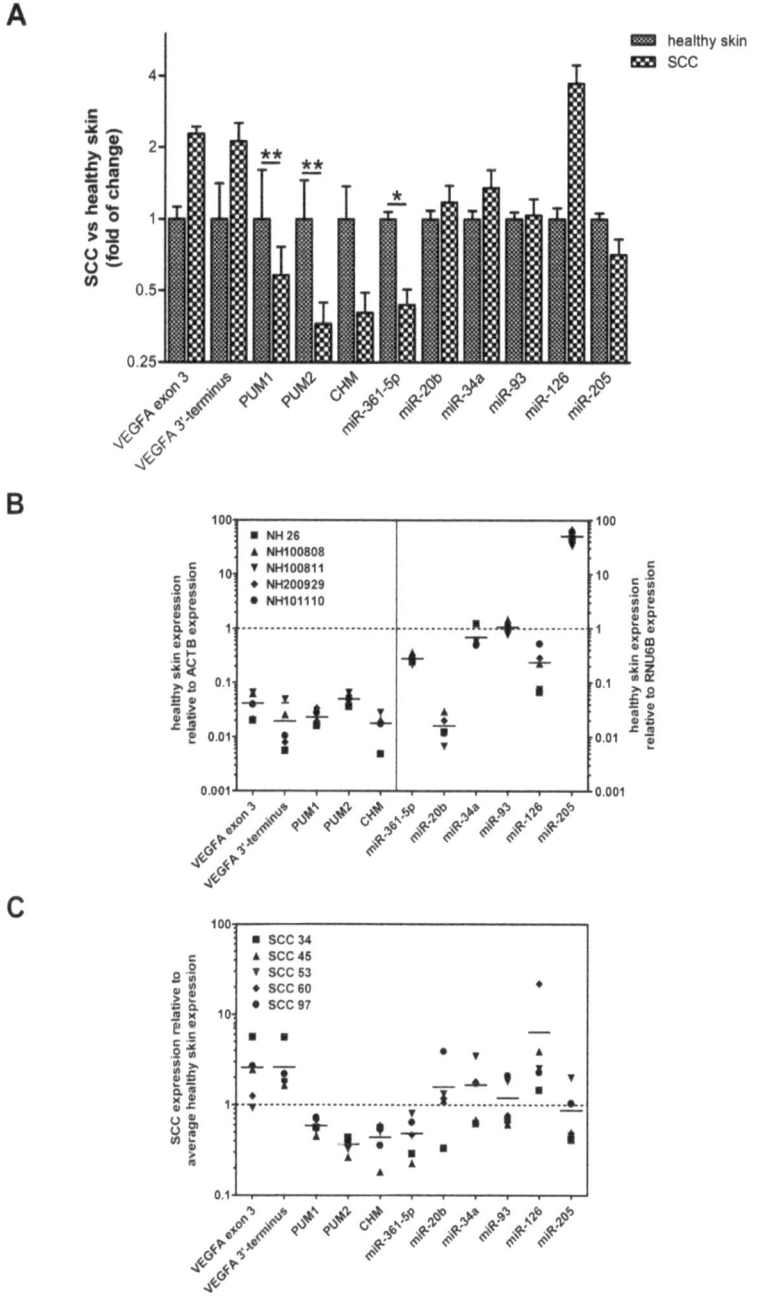

Figure 4.12 Relative changes in expression of selected mRNAs and mature miRNAs in healthy skin and cutaneous squamous cell carcinoma samples. For each group, five samples were analyzed by qRT-PCR, representing ten individuals. Experiments were performed in quadruplicates. Data are based on C_T values normalized to ACTB and RNU6B for mRNAs and miRNAs, respectively. (A) Fold differences in expression relative to the averages of expression in healthy skin samples ± S.D. are indicated for each assay. The Mann-Whitney U test was used to calculate P values (two-tailed; single and double asterisks denote P values <0.05 and <0.01, respectively). (B) For each of the indicated assays, the fold difference in expression with regard to the reference is plotted for each of the healthy skin samples. Horizontal bars represent means of all healthy skin samples. (C) For each assay and for each of the cutaneous squamous cell carcinoma samples, the fold difference in expression to the average of the healthy skin samples is indicated. Horizontal bars represent means of all squamous cell carcinoma samples. Note that data in (A to C) are plotted logarithmically.

assays, respectively). Notably, CHM did not correlate with VEGFA expression (r = 0.05 and 0.09, P = 0.4405 and 0.4014; Spearman's rank correlation, one-tailed; for the exon 3 and 3'-terminal assays, respectively).

Table 4.5 Correlation of microRNA 361-5p, Pum1 and Pum2 with VEGFA expression in healthy skin and squamous cell carcinoma samples. For each group, five samples were analyzed by qRT-PCR, representing ten individuals. Experiments were performed in quadruplicates. Data are based on C_T values normalized to ACTB and RNU6B for mRNAs and miRNAs, respectively. Based on the normalized expression levels, Spearman rank correlation coefficients (r) across all samples were calculated between VEGFA and the indicated messenger and miRNAs. P values (one-tailed) are indicated.

Assay	VEGFA exon 3		VEGFA 3'-terminus	
	r	P	r	P
VEGFA exon 3	1.00	n/a	0.83	0.0015
VEGFA 3'-terminus	0.83	0.0015	1.00	n/a
CHM	0.05	0.4405	0.09	0.4014
miR-361-5p	-0.58	0.0408	-0.60	0.0333
miR-20b	-0.05	0.4405	-0.22	0.2667
miR-34a	-0.08	0.4144	-0.43	0.1072
miR-93	-0.05	0.4405	-0.33	0.1733
miR-126	0.16	0.3257	-0.03	0.4669
miR-205	-0.26	0.2338	-0.44	0.1029
PUM1	-0.45	0.0935	-0.41	0.1221
PUM2	-0.30	0.2024	-0.16	0.3257

In summary, Pum1, Pum2 and miR-361-5p were significantly reduced in SCC expressing high levels of VEGFA, indicating that their dysregulation could contribute to the observed elevated VEGFA levels in SCC. Furthermore, we found that out of a panel of six miRNAs targeting VEGFA, only miR-361-5p levels were inversely correlated with VEGFA levels in the patient samples. Although we did not observe a similar inverse correlation between CHM and VEGFA mRNA levels, the correlation between CHM and miR-361-5p levels may be an indication that transcription of the miRNA precursor is dependent on *CHM* transcription.

4.3 Discussion

In the present study, we have used bioinformatics analysis tools to define potential recognition elements of the post-transcriptional repressors Pum1, Pum2 and miRNA-361-5p in a downstream conserved region of the human VEGFA 3'-UTR. By using luciferase reporter assays, we demonstrate that at least some of these elements possess regulatory potential, that elevated levels of the repressors negatively affect reporter activities, and that the repressive effects of the Pum proteins and miRNA-361-5p are consistent with an independent, additive model of combinatorial control. We have confirmed the repressive effect of miR-361-5p on VEGFA expression *in vitro* using ELISA. Importantly, we also found that the RNA levels of all three regulators are reduced in SCC samples compared to healthy skin. In the case of miR-361-5p, we could further show that its levels are inversely correlated with VEGFA expression in the patient samples. Taken together, we have identified three novel post-transcriptional regulators of VEGFA expression *in vitro*, and our results indicate that they may possibly affect cancer development or progression by modulating VEGFA expression in particular tumor types.

4.3.1 The putative Pum and microRNA recognition elements exhibit regulatory potential

The mutation of three critical nucleotides (Wang *et al.*, 2001, 2002) in each of the three putative PREs in the VEGFA 3'-UTR led to an increase in luciferase reporter activity, similar in extent to that of a 3'-UTR reporter of the known Pum target CDKN1B/p27 (Galgano *et al.*, 2008; Morris *et al.*, 2008; Kedde *et al.*, 2010) in which the two canonical

PREs were deleted. This de-repression is consistent with the usual role of PUF proteins as repressors of gene expression (see 4.1). While the effects may appear to be relatively moderate, it has to be considered that these are caused by the mutation or deletion of only 9 or 16 residues (VEGFA and CDKN1B reporter, respectively) out of more than a thousand each. While we did not discern in detail the contribution of the individual PREs, we have observed that the mutation of the most downstream PRE, PRE3, in the VEGFA 3'-UTR alone caused a similar response (all in the range of a 20-25% increase in relative *Renilla* activity) compared to the mutation of two (PRE2 and PRE3) or all three PREs (data not shown), suggesting that PRE3 is likely responsible for the majority of the changes. Likewise, a deletion of three residues in the 'seed' sequence confirmed a regulatory potential for the putative miR-361-5p MRE. Interestingly, and in contrast to the observations for the CDKN1B reporter with deletions of the two miR-221 seed sequences, this mutation caused a moderate decrease in relative luciferase activity. This is surprising, as we had expected that the mutation would negatively affect binding of the miRNA and thus, consistent with the well-established role of miRNAs as repressors of gene expression (see 2.2.2.2) and similar to the effects observed for the mutation of the putative PREs, lead to de-repression. There are, however, a number of reports in which miRNAs appear to upregulate the expression of targets, either directly or indirectly, depending on RNP context (reviewed in Vasudevan, 2011). This observation may thus well be a hint that the putative miR-361-5p MRE may possibly be under the cooperative control of two or more *trans*-acting factors. In this regard it may be important to note that PRE3 overlaps with the putative MRE of miR-361-5p (see Figure 4.2 A).

Our data indicate a regulatory potential of the putative PREs, at least for PRE3, as well for the predicted miR-361-5p MRE. The study of the individual contribution of each putative PRE to the total regulatory impact as well as the observations regarding the apparent activating potential of the putative miR-361-5p MRE may constitute entry points for future studies, which is further outlined in the following paragraphs.

4.3.2 Combinatorial control of VEGFA expression by microRNA 361-5p and the Pum proteins

Co-transfection of cells with CDKN1B and VEGFA 3'-UTR reporters and Pre-miR constructs of miR-221, and -361-5p respectively, led to a reduction in relative *Renilla* activities compared to the co-transfection with Pre-miR-control, thus suggesting that the putative miR-361-5p MRE in the VEGFA 3'-UTR is targeted by the respective miRNAs, just like the miR-221 MRE in the CDKN1B 3'-UTR (Galardi *et al.*, 2007; le Sage *et al.*, 2007). This is supported by the findings that mutation of the respective MREs abolished the repressive effects of elevated miR-221 and miR-361-5p levels. The absence of a complete de-repression using the VEGFA MRE mutant reporter – in contrast to the CDKN1B MRE double mutant reporter – could be explained by the relatively mild mutation (three/two nucleotides deleted in the MRE/seed sequence) of the former compared to the latter (complete deletion of both seeds), which consequently might allow residual binding of miR-361-5p. In contrast to the observations regarding the MRE-mutated VEGFA 3'-UTR reporter alone, these data suggest a repressive role of miR-361-5p on VEGFA expression. Interestingly, mutation of the PREs for both the CDKN1B and VEGFA reporters enhanced the repressive effect, adding further circumstantial evidence to the possibility of cooperative regulation between miR-361-5p and Pum, such as has been reported for CDKN1B (Kedde *et al.*, 2010).

Overexpression of both Pum1 and Pum2 in cells transfected with the VEGFA luciferase reporters led to a considerable decrease in relative *Renilla* activities compared to cells overexpressing a control protein, consistent with the widely established roles of PUF proteins as post-transcriptional repressors of gene expression (see 4.1). Together with the observed increases in luciferase activities of cells expressing a luciferase reporter with mutated canonical PREs, these data indicate that both Pum1 and Pum2 may likely contribute to the regulation of VEGFA expression. However, mutation of all three putative PREs did not fully abolish the observed repressive effect of Pum1 and Pum2 overexpression. While PUF proteins exhibit a certain amount of flexibility in the recognition of their target sequences (Lu and Hall, 2011; reviewed in Miller and Olivas, 2011), and the identified PREs of human Pum2 do not always fully match the consensus UGUAnAUA sequence (Hafner *et al.*, 2010b), the possibility that the mutated PREs allow residual binding of the Pum proteins seems unlikely, as the mutations cover the whole UGU triplet that is regarded critical for the target recognition of PUF proteins (Wang *et al.*, 2001, 2002) and thus probably does not allow such profound mutations. The failure of PRE mutant reporters to effect a complete de-repression is thus more likely to stem from the presence of additional PREs that mediate binding to Pum1 and Pum2 even in the absence of the canonical PREs. This is consistent with the observations that the complete deletion of both consensus PREs in the CDKN1B 3'-UTR is also unable to effect a de-repression, although the functionality of at least the upstream PRE has been proven (Kedde *et al.*, 2010). Furthermore, the global identification of Pum2 PREs in Flp-In-293-derived cells using PAR-CLIP (Hafner *et al.*, 2010b) supports the idea of non-canonical PREs: While two likely binding sites for Pum2 were indeed found in the VEGFA 3'-UTR, the corresponding sequence tags are located approximately 60 nt upstream of PRE2, and downstream PRE3 respectively (see Figure 4.1), and they contain no PRE consensus motifs or UGU triplets. Instead, both contain UCUACAUA sequences, which differ from the canonical PRE motif in only one nucleotide and could thus be likely candidates for such 'cryptic' PRE motifs. The absence of sequence tags covering the putative consensus PREs, and particularly PRE3, might be explained by the masking of functional PREs by other *trans*-acting factors, for example miR-361-5p, as well as technical limitations, and thus does not necessarily mean that these sites are not amenable to regulation by Pum1 and/or Pum2. *In vitro* binding assays conducted in our lab support this explanation, as a transcript not including PRE2, PRE3 and either of the sequence tags was nevertheless bound by the Pum1 and Pum2 homology domains (Galgano et al., 2008).

Taken together, these data indicate that VEGFA expression may be under the combinatorial control of Pum1, Pum2 and miR-361-5p, which all act as repressors on its 3'-UTR in our *in vitro* reporter assays. While the *cis*-regulatory element for miR-361-5p could be identified by mutational studies, the PRE identity is less clear: While PRE3 appears to possess at least some regulatory potential, the VEGFA 3'-UTR likely contains additional, non-canonical PREs. Moreover, other potential Pum binding sites, such as the consensus motifs PRE1 and PRE2, may possibly gain functionality only when other PREs are blocked or mutated, thus making the study of their individual role a complex task.

4.3.3 The regulation of VEGFA expression by microRNA 361-5p and the Pum proteins may be dependent on each other

The data from experiments in which we simultaneously increased the levels of miR-361-5p and either of the Pum proteins largely indicate additive effects of the two classes of repressors. This would suggest the absence of cooperativity, which is in contrast to some of the previously discussed results. Moreover, due to the overlap of the miR-361-5p MRE and PRE3, it appears unlikely that no competition exists between these motifs, if PRE3 indeed represents a functional Pum recognition element, as is supported by the increased relative *Renilla* activity when comparing the PRE triple mutant with the wild type VEGFA reporter. There are several possible reasons why we failed to pick up cooperativity in our experimental setup. First of all, the repression exerted on the VEGFA 3'-UTR by both the Pum proteins and miR-361-5p is relatively moderate, and it is reasonable to assume that the modulation of the effects by possible cooperativity may be even smaller in extent. But the sensitivity and robustness of our assay system is limited, so that it is possible that we would not be able to reliably detect such differences. This scenario is supported by the data obtained from the PRE mutant reporter, in which deviations from an additive model of up to 20% in either direction were observed. However, due to the sample variation, these did not reach significance. On the other hand, increasing the levels of both types of regulators may not be the ideal setup to discover competitive effects between two repressors, as their relative ratios may not be affected. The synergy between Pum1- and miR-221-mediated regulation of the CDKN1B 3'-UTR, which has previously been established (Kedde *et al.*, 2010), could also not be confirmed using our experimental strategy. In accordance with the mentioned study, the effect of depleting either or even both of the repressors should thus be tested in future experiments. The aforementioned possibility of the existence of alternative PREs, whether active in itself or requiring the blockage of other PREs to gain functionality, may be another explanation for the inability to detect cooperativity, as the effects of potential competition between the regulators for a particular PRE, e.g. PRE3, may be diluted by the use of distant alternative, perhaps non-canonical PREs. Finally, it is conceivable that crosstalk between the regulators requires the presence or absence of additional factors to become visible. The large set of known post-transcriptional regulators of VEGFA expression (see 2.4.1), as well as the extremely high density of predicted MREs in the downstream conserved region of the VEGFA 3'-UTR (0.41 MREs/nucleotide, compared to 0.29 and 0.14 for the upstream conserved and non-conserved regions, respectively; see Table 4.1 and 7.4) indeed favor a complex picture of combinatorial control, with many players being involved. This implies the potential for a high degree of competition or other cooperative effects between miRNAs, or between miRNAs and RNA-binding proteins. Indeed, cooperativity has already been observed between other miRNAs regulating VEGF expression *in vitro* (Hua *et al.*, 2006), as well as between miRNAs and RNA-binding proteins (Jafarifar *et al.*, 2011).

Considering that the decrease of reporter activity in response to increased repressor levels appeared to be additive, the results suggest the absence of cooperativity under the tested conditions and instead imply a model of independent control mechanisms for miR-361-5p on the one hand and the Pum proteins on the other. But considering all of the results from our *in vitro* reporter assays, a definite conclusion regarding the nature of the combinatorial control between the two repressors cannot be drawn. The use of reporters with a limited 3'-UTR context and number of potential PREs (e.g. the region between PRE2 and the first potential downstream UCUACAUA Pum binding site; Hafner *et al.*, 2010b) might greatly facilitate the dissection of the *cis*-regulatory elements, because it would (a) restrict the number of potential alternative Pum binding sites and other *cis*-regulatory elements, and (b) likely lead to stronger effects and thus a better assay sensitivity. While the biological significance of results obtained from such short, "out-of-context" reporters may not be immediately clear, or even misleading, they could still serve as a starting point for a subsequent in-depth functional analysis.

4.3.4 The influence of microRNA 361-5p and Pum1 on VEGFA secretion rates

Consistent with findings indicating that the regulatory impact of altered miRNA levels on endogenous protein levels of targets is often weak (Baek *et al.*, 2008; Selbach *et al.*, 2008), we have also observed mild inhibitory effects of miRNA 361-5p on VEGFA secretion rates in HaCaT and A431 cells. A difference in VEGFA miR-361-5p MRE occupancy by endogenous miR-361-5p between the two cell lines could potentially explain the differential effects of miR-361-5p mimic or antisense inhibitor on their VEGFA secretion rates: A high occupancy ('saturation') of the MRE in HaCaT cells, which express similar miR-361-5p but lower VEGFA levels compared to A431, could account for the absence of a significant decrease in VEGFA secretion upon the addition of miR-361-5p mimic. Conversely, a low occupancy of the MRE in A431 cells could render the cells unresponsive to miR-361-5p inhibition.

However, in this simple scenario the impact of increased miR-361-5p levels in A431 and miR-361-5p inhibition in HaCaT cells should be proportional to the amount of added miRNA mimic or antisense inhibitor. The absence of such a miRNA dose-response could be a further indication that one or more additional factors differentially modulate miR-361-5p activity in the two cell lines, systemically or specifically, by influencing the availability, accessibility or functionality of the miRNA or its recognition element.

The influence of Pum1 overexpression and knockdown on VEGFA expression in A431 cells was previously assessed in our laboratory (Galgano, 2010). Using ELISA, it was shown that knockdown of Pum1 led to a significant decrease in VEGFA secretion rates and intracellular VEGFA protein levels. Conversely, Pum1 overexpression resulted in a significant increase in intracellular VEGFA, while a similar effect on secreted levels did not reach significance. These data add further evidence for a regulatory role of Pum proteins on VEGFA expression and indicate that the nature, impact, and direction of the regulation are strongly dependent on the cellular context and thus likely on the crosstalk with other post-transcriptional regulators.

4.3.5 A potential role of microRNA 361-5p and Pum proteins in cancer development and progression

We have shown that levels of miR-361-5p, but not those of the known VEGFA-regulating miRNAs miR-20b, -34a, -93, -126, and -205, inversely correlate with VEGFA expression in SCC compared to healthy skin samples, corresponding to previously reported findings (Dziunycz *et al.*, 2010). Moreover, Pum1 and Pum2 mRNA as well as miR-361-5p levels were significantly reduced in the SCC samples, suggesting that any or all of these repressors may contribute to elevated VEGFA levels at least in this type of cancer. Studies with larger sample numbers and follow-up data should establish the value of these regulators as diagnostic and prognostic markers or potential drug targets.

These findings further underline the presence of cancer-specific miRNA profiles (reviewed in Calin and Croce, 2006), the emerging and possibly widespread role of RNA-binding proteins in disease in general (reviewed in Lukong *et al.*, 2008), and the interaction of the two classes of post-transcriptional gene regulators in cancers specifically (reviewed in van Kouwenhove *et al.*, 2011). The application of methods for the unraveling of post-transcriptional regulatory networks, as well as expression profiling of cancers with respect to miRNAs, RBPs and other post-transcriptional regulators, should allow us to gain further insights into these intricate and exciting new mechanisms. Eventually, the integration of the available knowledge on transcriptional, post-transcriptional, post-translational, epigenetic, and other types of gene regulation, should greatly broaden our understanding of the dysregulation of tumor suppressors, oncogenes and critical players in other diseases.

Interestingly, the data from SCC and healthy skin samples add another possibility for potential crosstalk between post-transcriptional regulatory mechanisms. The localization of the putative recognition elements for Pum1/2 and miR-361-5p close to the mRNA's 3'-terminus may render it prone to inaccessibility due to degradation or alternative polyadenylation. Regarding the latter, it has recently been proposed that proliferating cells may employ alternative polyadenylation to shorten the 3'-UTRs of transcripts and thus escape miRNA- and RNA-binding protein-mediated post-transcriptional repression (Sandberg *et al.*, 2008). Indeed, it was found that the VEGFA transcript uses two different polyadenylation sites in mice, although no differential usage of the signals was observed between normoxic and hypoxic conditions (Dibbens *et al.*, 2001). Similarly, our data from SCC and healthy skin samples indicate an apparent difference in expression levels between the coding region and the downstream region of the 3'-UTR within but not in between both sample groups, suggesting that alternative polyadenylation may indeed limit the availability of the miR-361-5p MRE (Figure 4.12 B). Further studies into the differential regulation of VEGFA 3'-UTR length, the interplay between miRNAs, as well as the crosstalk with other post-transcriptional regulators may shed light on these issues.

4.3.6 Bioinformatics analyses suggest common functions of microRNA 361-5p and Pum proteins beyond the regulation of VEGFA expression

It has been proposed that miRNAs often act as master regulators of PTGR (reviewed in Kanitz and Gerber, 2010) and thus could well target multiple targets within the same pathway. Indeed, besides regulating the expression of VEGFA, miR-361-5p, Pum1 and Pum2 are predicted or were found to target thousands of different mRNAs, and the proteins encoded by these often act in similar pathways. Strikingly, out of all pathways significantly enriched among the genes (putatively) targeted by either regulator, around 45% are enriched among genes predicted to be targeted by miR-361-5p and found to be associated with either Pum1 or Pum2. In addition to many others, the VEGF pathway, as well as angiogenesis in general were among the enriched pathways, consistent with the important roles that have been proposed for RNA-binding proteins and miRNAs in the regulation of angiogenesis (reviewed in Chang and Hla, 2011). Even though the reliability of prediction software and GO annotation is often questionable at the level of individual targets, the large number of commonly enriched pathways suggests a large potential for common functions, possibly extending those suggested by their identification as repressors of VEGFA expression. Future studies should establish the biological relevance of these predictions using functional assays or *in vivo* experiments.

4.3.7 Conclusion

We were able to demonstrate the validity of our combined evidence- and prediction-based approach for the identification of regulators of the expression of VEGFA, whose dysregulation is implicated in several human malignancies. However, the downsides of such a biased approach lie in the difficulties associated with the unambiguous identification of *cis*-regulatory elements. Furthermore, the study of combinatorial control mechanisms may be severely hampered by the complexity of PTGR programs and the high interconnectivity of the underlying networks. This is particularly problematic for messages like VEGFA whose expression is extensively controlled at the post-transcriptional level, both by RNA-binding proteins and miRNAs. Nevertheless, we were able to identify three novel *trans*-acting factors, the miRNA miR-361-5p and the RNA-binding proteins Pum1 and Pum2, which are able to repress VEGFA expression *in vitro*. We could further establish that all regulators exhibited reduced expression in cutaneous squamous cell carcinoma samples expressing high levels of VEGFA, indicating that the regulation demonstrated *in vitro* may have medical implications.

4.4 Materials and Methods

4.4.1 Ethics statement

The collection of specimens from clinically indicated excisions for this study was explicitly approved by the institutional review board (Kantonale Ethikkommission Zürich). Informed consent (both written and verbal) was obtained from patients for the use of their skin samples in this research project.

4.4.2 Plasmids

The wild type and mutated versions of the CDKN1B/p27 3'-UTR (nt 1070-2403 of

RefSeq mRNA NM_004064.3 at GenBank) were amplified from pGL3-*CDKN1B*-3'UTR, pGL3-*CDKN1B*-3'UTR-PRE-DM, and pGL3-*CDKN1B*-3'UTR-MRE-DM respectively (kindly provided by Martijn Kedde, NKI Amsterdam), using the primers *CDKN1B*-3'-UTR-fwd and *CDKN1B*-3'-UTR-rev (containing XhoI and NotI restriction sites, respectively; see 7.1.1). A fragment comprising nucleotides 43,753,225 to 43,754,253 of human chromosome 6 (Build GRCh37/hg19, February 2009), containing nucleotides 926 to 1925 of the 3'-UTR of human VEGFA (isoform a, NM_001025366.2), was amplified from HeLa S3 genomic DNA using KOD Hot Start DNA Polymerase (Novagen) and the primers *VEGFA*-3'-UTR-fwd and *VEGFA*-3'-UTR-rev (containing XhoI and NotI restriction sites, respectively; see 7.1.1). Amplicons were purified with the QIAquick PCR Purification kit (QIAGEN, Cat. No. 28104) and inserted into pCR-Blunt II-TOPO (Invitrogen), using the Zero Blunt TOPO PCR Cloning Kit (Invitrogen, Cat. No. K2830-20) according to the manufacturer's recommendations. The resulting plasmids, pCR-Blunt II-TOPO-*CDKN1B*-3'-UTR, pCR-Blunt II-TOPO-*CDKN1B*-3'-UTR-PRE-DM, pCR-Blunt II-TOPO-*CDKN1B*-3'-UTR-MRE-DM, and pCR-Blunt II-TOPO-*VEGFA*-3'-UTR, were verified by sequencing of the inserted regions. Mutations in the putative Pum and miR-361-5p recognition elements of the VEGFA 3'-UTR were introduced using the QuikChange I site-directed mutagenesis kit (Stratagene) according to the manufacturer's recommendations. PREs were sequentially mutated using the following oligonucleotide pairs in the indicated order: VEGFA-PRE3-MUT-fwd/-rev (PRE3 mutation), VEGFA-PRE2/3-MUT-fwd/-rev (PRE3/2 double mutation), VEGFA-PRE1-MUT-fwd/-rev (PRE3/2/1 triple mutation). The VEGFA-MRE-MUT-fwd/-rev oligonucleotide pair was used to mutate the MRE in pCR-Blunt II-TOPO-*VEGFA*-3'-UTR. See 7.1.2 for oligonucleotide sequences. All mutations were verified by sequencing of the appropriate regions. To generate the luciferase reporter constructs psiCHECK-2-*CDKN1B*-3'-UTR ('wt CDKN1B'), psiCHECK-2-*CDKN1B*-3'-UTR-PRE-DM ('CDKN1B-PRE-DM'), psiCHECK-2-*CDKN1B*-3'-UTR-MRE-DM ('CDKN1B-MRE-DM'), psiCHECK-2-*VEGFA*-3'-UTR ('wt VEGFA'), psiCHECK-2-*VEGFA*-3'-UTR-PRE-TM ('VEGFA-PRE-TM'), and psiCHECK-2-*VEGFA*-3'-UTR ('VEGFA-MRE-MUT'), wild type and mutated *CDKN1B* and *VEGFA* 3'-UTR fragments were then subcloned into psiCHECK-2 (Promega) via *Xho*I and *Not*I restriction sites.

The coding sequence for Pum1 (corresponding GenBank entry CV027786.1) was from pDONR223-Pum1 (Thermo Scientific Inc., Cat. No. OHS1770-93822323; kindly provided by Alexander Wepf, Institute of Molecular and Systems Biology, ETH Zurich). The coding sequence for Pum2 (corresponding to GenBank entry BC143550.1) was amplified from pCMV6-XL5-Pum2 (OriGene, Cat. No. SC112640) with primers Pum2-CDS-attB-fwd and Pum2-CDS-attB-rev (both containing attB recombination sites; see 7.1.1). The amplicon was purified with the QIAquick PCR Purification kit (QIAGEN, Cat. No. 28104) and inserted into pDONR221 (Invitrogen, 12536-017) by recombination, using BP Clonase II Enzyme Mix (Invitrogen, Cat. No. 11789-020) according to the manufacturer's recommendations. The Pum2 coding sequence of the resulting plasmid, pDONR221-Pum2, was verified by sequencing. To generate pTO-HA-Strep-GW-FRT-Pum1 and pTO-HA-Strep-GW-FRT-Pum2, pTO-HA-Strep-GW-FRT (kindly provided by Alexander Wepf, Institute of Molecular and Systems Biology, ETH Zurich; see 7.7) was recombined with pDONR-223-Pum1 and pDONR221-Pum2, respectively, using LR Clonase Enzyme Mix (Invitrogen; Cat. No. 11791-020) according to the manufacturer's recommendations.

4.4.3 Cell culture and tissue samples

All cell lines were cultured in Dulbecco's modified Eagle's medium (DMEM; Invitrogen, Cat. No. 41966) supplemented with 10% fetal bovine serum (Invitrogen, Cat. No. 10270-106) and 1x antibiotic-antimycotic (Invitrogen, Cat. No. 15240-062) in 5% CO_2 at

37°C. Flp-In-293 cells were cultured in the presence of 1 mg/mL zeocin (Invitrogen, Cat. No. R250-01), Flp-In-293-Pum1/2 and -eGFP cells in the presence of 200 µg/mL hygromycin B (Invitrogen, Cat. No. 10687-010).

Flp-In-293 (Invitrogen, Cat. No. R750-07) and Flp-In-293-eGFP cell lines were kindly provided by Alexander Wepf (Institute of Molecular and Systems Biology, ETH Zurich). HEK293 (Graham *et al.*, 1977) and A431 (Giard *et al.*, 1973) cells were purchased from ATCC (CRL-1573 and CRL-1555, respectively). HaCaT (Boukamp *et al.*, 1988) cells were obtained from Cell Lines Service (Cat. No. 300493). Flp-In-Pum1 and Flp-In-Pum2 cells were generated by co-transfection of 500,000 Flp-In-293 cells with 720 ng pOG44 (Invitrogen, Cat. No. V6005-20; kindly provided by Alexander Wepf, Institute of Molecular and Systems Biology, ETH Zurich) and 80 ng of either pTO-HA-Strep-GW-FRT-Pum1 or -Pum2, respectively, using the FuGENE HD Transfection Reagent (Roche, Cat. No. 04883560001) according to the manufacturer's recommendations. Stably transformed cells were selected by supplementing the medium with hygromycin B (200 µg/mL; Invitrogen, Cat. No. 10687-010).

Squamous cell carcinoma (SCC) samples were obtained at the time of surgery. Normal skin was obtained from abdominoplastic reductive surgery. All specimens' diagnoses were confirmed by a board-certified dermatohistopathologist. Four mm punch biopsies from SCC or normal skin were placed in preheated phosphate-buffered saline (PBS; Invitrogen, Cat. No. 14190) at 60°C for 45 seconds, and then chilled on ice in 0.1% PBS for one minute, followed by mechanical separation of epidermis and dermis by scratching. The epidermis was homogenized in TRIzol reagent (Invitrogen) and stored at -80°C. RNA was extracted according to the manufacturer's recommendations. Quantity and quality of extracted RNA was assessed by spectrophotometry with a NanoDrop 1000 (Thermo Fisher Scientific Inc.) and a 2100 Bioanalyzer (Agilent Technologies), respectively. All RNA samples had an RNA

Integrity Number (RIN) of higher than 7.0.

4.4.4 MicroRNA target gene prediction and pathway analysis

Predictions of human miRNA targets were downloaded from microRNA.org (Betel et al., 2010) (August 2010 release, http://www.microrna.org/, accessed: August 12th, 2011; all predictions with mirSVR of less than -0.1 were considered), TargetScan (Friedman et al., 2009) (Release 5.2; http://www.targetscan.org/, accessed: August 12th, 2011; all predicted miRNA recognition elements were considered, regardless of conservation), DIANA-microT v3.0 (Maragkakis et al., 2009) (Release 3.0, http://diana.cslab.ece.ntua.gr/microT/, accessed: August 12th, 2011), miRDB (Wang, 2008) (Release 3.0, http://mirdb.org/miRDB/, accessed: August 12th, 2011), and MicroCosm (Griffiths-Jones et al., 2008) (Release 5, www.ebi.ac.uk/enright-srv/microcosm/, accessed: August 12th, 2011). The miR-361-5p sequence was obtained from miRBase (Kozomara and Griffiths-Jones, 2011) (Release 17, http://www.mirbase.org/, accessed: May 5th, 2011). RNAhybrid (Rehmsmeier et al., 2004) (http://bibiserv.techfak.uni-bielefeld.de/rnahybrid/) was used online, with default settings and the following sequences: UUAUCAGAAUCUCCAGGGGUAC (miR-361-5p, miRNA) and UGUAUAUAUGTGAUUCUGAUAAA (VEGFA 3'-UTR fragment containing the putative miR-361-5p MRE, target RNA). For the pathway analysis, predicted targets for miR-361-5p and experimentally verified Pum1 and Pum2 targets (Morris et al., 2008; Galgano et al., 2008; Hafner et al., 2010b) were converted to Entrez identifiers using DAVID (Huang et al., 2008) (http://david.abcc.ncifcrf.gov/), if not present in the respective outputs. Results for each group were pooled, filtered for unique records, and subjected to gene set enrichment analysis with PANTHER (Thomas et al., 2006) (http://www.pantherdb.org/tools/compareToRefList Form.jsp). For VEGFA pathway analysis, the KEGG PATHWAY database (Kanehisa et al., 2010) (http://www.genome.jp/kegg/pathway.html) was consulted. All services were used with default settings.

4.4.5 Immunoblot analysis

Cells were treated with tetracyclin (2 µg/mL; 24 h) and lysed with RIPA buffer (Cell Signaling Technology, Cat. No. 9806S) supplemented with 1 mM phenylmethanesulfonyl fluoride (Sigma-Aldrich, Cat. No. P7626) according to the manufacturer's recommendations. Protein concentrations were determined using the Bio-Rad Protein Assay according to the manufacturer's recommendations and spectrophotometric analysis with a NanoDrop 1000 (Thermo Fischer Scientific, Inc.) at a wavelength of 595 nm. 20 µg protein extracts were supplemented with NuPAGE LDS Sample Buffer (Invitrogen, Cat. No. NP0008) and NuPAGE Sample Reducing Agent (Invitrogen, Cat. No. NP0009), loaded on NuPAGE Novex 4-12% Bis-Tris gels (1mm, 10 wells; Invitrogen, Cat. No. NP0321BOX). Proteins were separated by polyacrylamide gel electrophoresis in NuPAGE MOPS SDS Running Buffer (Invitrogen; Cat. No. NP0001) supplemented with NuPAGE Antioxidant (Invitrogen; Cat. No. NP0005), according to the manufacturer's recommendations. Proteins were transferred onto nitrocellulose membranes (Bio-Rad, Cat. No. 162-0115) in a Mini-PROTEAN II Electrophoresis Cell (Bio-Rad) in the presence of transfer buffer (25 mM Tris, 192 mM glycine, 20% methanol; 200 mA per membrane, 1 h, 4°C). Membranes were blocked with 5% low-fat milk powder in PBS (1 h, RT, 300 rpm). Anti-HA antibody (clone HA-7; Sigma-Aldrich, Cat. No. H3663-200UL) was added at 0.5 µg/mL and the incubation was continued for an additional hour. Membranes were washed twice with PBS containing 0.05% Tween-20 (Sigma-Aldrich, Cat. No. P2287-100ML) for 5 min each (300 rpm) and incubated (1 h, RT, 300 rpm) in PBS containing 5% low-fat milk powder and a 1:5,000 dilution of a horseradish peroxidase-coupled anti-mouse antibody (Amersham, Cat. No. NA931V). Membranes were washed four times as described above, briefly dried and developed with the ECL Plus

Western Blotting Detection System (Amersham, Cat. No. RPN2132) according to the manufacturer's recommendations. Bands were visualized with a Bio-Rad Universal Hood II.

4.4.6 Flow cytometry

To assess the efficiency of small RNA transfections, 125,000 A431, HaCaT or HEK293 cells were reverse transfected with 10, 30 or 100 nM Cy3 dye-labeled Pre- or Anti-miR Negative Control #1 (Applied Biosystems; see 7.2), or mock-transfected, using siPORT NeoFX (Applied Biosystems) according to the manufacturer's recommendations. Transfections were performed in triplicate in 24-well plates. For the analysis of eGFP expression, Flp-In-293-eGFP cells were treated with the indicated concentrations of tetracyclin. In both cases, cells were detached with 0.05% trypsin-EDTA (Invitrogen, Cat. No. 25300-054) after 24 h and washed with PBS. Dye-labeled small RNAs or eGFP were excited with a blue laser (excitation wavelength = 488 nm) and analyzed with a FACScan flow cytometer (Becton Dickinson). At least 5,000 or 50,000 events were recorded for each sample for the analysis of small RNA transfections and eGFP levels, respectively. Data were analyzed with WinMDI 2.8.

4.4.7 Quantitative reverse transcription PCR

For each reaction, cDNA was prepared from 10 ng total RNA using the TaqMan MicroRNA Reverse Transcription Kit (Applied Biosystems) for miRNA detection, or the High Capacity cDNA Reverse Transcription Kit (Applied Biosystems) for mRNA detection. miRNA and gene expression assays were purchased from QIAGEN (SYBR; ACTB, PUM1, PUM2) or Applied Biosystems (TaqMan; all others). See 7.3 for an overview of commercial qRT-PCR assays. Quantitative PCR reactions were performed in quadruplicates, using FastStart TaqMan Probe Master (Rox; Roche) or FastStart Universal SYBR Green Master

(Rox; Roche, Cat. No. 04913914001) in an AB 7900 HT Fast Real-Time PCR System (Applied Biosystems). Quantification was performed using the $2^{-\Delta\Delta CT}$ method (Schmittgen and Livak, 2008), with RNU6B and ACTB serving as references for the normalization of miRNA and mRNA expression levels, respectively. Equal amplification efficiencies of near 100% were assumed for all assays, based on the manufacturers' assertions.

4.4.8 Immunocytochemistry

The chambers of an 8-chamber slide (BD Falcon, Cat. No. 354108) were coated with 5 µg/mL fibronectin (Sigma-Aldrich, Cat. No. F0895) diluted in PBS for 20 min at RT. 250 µL of an Flp-In-293-Pum1 cell suspension (8×10^4 cells/mL in DMEM supplemented with 10% fetal bovine serum and 1x antibiotic-antimycotic) were seeded into each of the chambers. 16 h later, medium was replaced with fresh medium containing 1 µg/mL tetracyclin (2 mg/mL stock solution in EtOH). 23 hours later, sodium arsenite (0.05 M; Sigma, Cat. No. 35000-1L-R) was added to the medium at 0.5 mM. After 45 min, cells were fixed (4% paraformaldehyde in PBS; 15 min, RT, 300 rpm) and permeabilized (0.1% Triton X-100, 0.5% BSA, 1 µg/mL Hoechst 33342 in PBS; 3 min, RT). After pre-incubation in blocking buffer (0.5% BSA, 5% donkey serum in PBS) for 30 min at RT, cells were co-stained with an anti- Pum1 (25 µg/mL; A300-201A, Bethyl Laboratories, Inc., Cat. No A300-201A) and an anti-HA antibody (clone HA-7; 2 µg/mL; Sigma, Cat. No. H3663) in blocking buffer (1 h 30 min, 300 rpm). Cells were washed twice in PBS with 0.02% Tween-20 (Sigma-Aldrich, Cat. No. P2287-100ML) for 5 min each (300 rpm), and then incubated with 1:200 dilutions of Alexa488-coupled anti-goat and Alexa594-coupled anti-mouse antibodies (both from Molecular Probes, Cat. Nos. A-11055 and A-21203, respectively) in blocking buffer (1 h, RT, 300 rpm) in the dark. Cells were washed three times as described above and mounted in Mowiol 4-88 (Calbiochem, Cat. No. 475904). Images of the cells were acquired with a Leica TCS SP2 confocal microscope

(Light Microscopy Center, ETH Zurich). Overlay images were created with Adobe Photoshop CS5.

4.4.9 Luciferase reporter assays

20,000 of the indicated cells were reverse transfected with 20 ng of the indicated luciferase reporter constructs using polyethylenimine (Polysciences, Inc., Cat. No. 23966). For complex formation, DNA and polyethylenimine stock solution (1 mg/mL in water) were diluted in Opti-MEM I (Invitrogen, Cat. No. 51985-026), to 20 µg/mL and 60 µg/mL, respectively. Both solutions were incubated for 10 min at room temperature, and then mixed at equal volumes (mass ratio polyethylenimine to DNA = 3:1; final polyethylenimine concentration = 30 µg/mL). After incubation for 20 min at room temperature, solutions were diluted 1:10 in DMEM (Invitrogen, Cat. No. 41966) to achieve a final DNA concentration of 1 µg/mL. 20 µL of the transfection mixes were added to each well of a 96-well plate, followed by the addition of 80 µL of cell suspension in DMEM (2.5×10^5 cells/mL). Where applicable, 16 hours after transfection of the luciferase reporter plasmids, the cells were further transfected with 50 nM of Pre-miR-361-5p or Pre-miR Negative Control #1 (Pre-miR-control; Applied Biosystems; see 7.2) by using siPORT NeoFX (Applied Biosystems) according to the manufacturer's recommendations. All transfections were performed in triplicate. Where applicable, the medium was replaced with fresh medium containing 1 µg/mL tetracyclin (2 mg/mL stock in EtOH) or EtOH only after 24 hours. In all cases, the medium was aspirated 64 hours after the initial transfection. Cells were lysed with a mixture of 15 µL Luciferase Assay Reagent II (Promega) and 15 µL nuclease-free water (Invitrogen, Cat. No. 10977). Firefly luciferase activity was measured after 10 min. Subsequently, 15 µL Stop & Glo Reagent (Promega) were added and *Renilla* luciferase activity was measured after 10 min. Luciferase activity measurements were performed in an LMAX II 384 luminometer

(Molecular Devices) with 5 seconds integration time. For each triplicate, the mean *Renilla*/firefly ratio was calculated.

4.4.10 Enzyme-linked immunosorbent assay

20,000 A431 or HaCaT cells were reverse transfected with 10, 30 or 100 nM Pre- or Anti-miR-361-5p or Pre-/Anti-miR Negative Control #1 (Applied Biosystems; see 7.2) with siPORT NeoFX (Applied Biosystems) according to the manufacturer's recommendations. Transfections were performed in triplicate in 96-well plates. 24 hours after transfection, supernatants were collected and centrifuged to remove cell debris (1000 g for 3 min at room temperature). VEGFA protein levels were determined using the Human VEGF-A Platinum ELISA kit (eBioscience, Cat. No. BMS277) according to the manufacturer's recommendations. After subtraction of blank values, triplicates were averaged and quantified using a standard curve prepared from serial dilutions of purified VEGFA.

4.5 Contributions

André P. Gerber, Michael Detmar, Jochen Imig, Alessia Galgano, Jonathan Hall (Institute of Pharmaceutical Sciences, ETH Zurich), Günther F. L. Hofbauer, and Piotr J. Dziunycz (Department of Dermatology, University Hospital Zurich, Zurich) have contributed to the conception or design of experiments or provided helpful discussions. Jochen Imig, under the supervision of André P. Gerber and Jonathan Hall, performed the flow cytometry analysis of small RNA-transfected A431 and HaCaT cells. Piotr J. Dziunycz, under the supervision of Günther F. L. Hofbauer, provided the healthy skin and cutaneous squamous cell carcinoma RNA samples and performed the RNA quality control experiments. Michael Forrer (Institute of Pharmaceutical Sciences, ETH Zurich), under the supervision of Alessia Galgano and André P. Gerber, has generated the wild type CDKN1B luciferase reporter.

Alexander Kanitz, under the supervision of André P. Gerber and Michael Detmar, performed all other experiments. Fabienne Bereiter and Alexander Svensson (Institute of Pharmaceutical Sciences), under the supervision of Alexander Kanitz and André P. Gerber, have helped with the generation of mutant CDKN1B and VEGFA luciferase reporters. Sinem Karaman (Institute of Pharmaceutical Sciences), under the supervision of Michael Detmar, has helped with statistical analyses. Martijn Kedde, under the supervision of Reuven Agami (NKI, Amsterdam, Netherlands), and Alexander Wepf, under the supervision of Matthias Gstaiger (Institute of Molecular and Systems Biology), have contributed plasmids and cell lines as indicated in the main text.

5 A Novel RNA Tandem Affinity Tag for the Purification of Ribonucleoprotein Particles

5.1 Introduction

In eukaryotes, a large number of regulatory programs govern all major events in the entire life span of messenger and non-coding RNAs (see 2.1.1). The underlying 'code', in the form of *cis*-acting sequence and structural elements, is stored in the RNA itself (see 2.1.2), while interpretation and execution of the programs requires the help of additional protein and RNA factors acting in *trans* (see 2.1.3). Together, these proteins and RNAs form RNPs, highly dynamic macromolecular complexes that represent the functional units of PTGR (see 2.1.4). RNPs and their components are organized in intricate, densely meshed post-transcriptional gene regulatory networks (GRNs) that orchestrate and coordinate the execution of regulatory programs simultaneously, for the whole transcriptome, in the correct spatiotemporal context and in response to continuously changing environmental stimuli (see 2.2). In addition to complex loop circuits and cooperation, GRN architecture heavily relies on combinatorial control principles to accomplish this feat: A single *trans*-acting factor generally binds multiple RNAs (2.2.1.1); in turn, a single RNA is generally bound by multiple *trans*-acting factors (2.2.1.2). Thanks to a range of elegant affinity purification-based methodologies for RNP analysis that have been developed and continuously improved over the last decade or so, several streamlined protocols now exist for the identification of RNA species that are bound by a particular *trans*-acting (RNA-binding) protein (see 2.3.1).

Unfortunately, a widely applicable approach for the systematic identification of the *trans*-acting factors binding a particular RNA is currently unavailable. Such a methodology, i.e. one that allows the purification of RNPs via their RNAs, would enable us to directly and comprehensively examine the post-transcriptional combinatorial control exerted on a given

RNA. For the study of cancers, which arise from the dysregulation of a relatively small number of genes, such a 'gene-centered' ribonomics approach would therefore be highly desirable (see 2.4). Several attempts at establishing such a procedure have been made, but they generally suffer from lack of sensitivity, low efficiency or their artificial or complicated setup (see 2.3.2).

Based on the shortcomings of current approaches and the complexity of RNP dynamics, we believe that a powerful, yet versatile approach towards the affinity purification of RNA molecules and the identification of associated proteins and RNAs has to guarantee that RNP formation occurs unimpeded, in its native conformation and location inside the cell (i.e. *in situ*), and with as little disturbance to the cells as possible. Therefore it has to rely on structural determinants that can be expressed and purified directly from lysed cells. Apart from escaping the additional, potential interference with normal RNP formation caused by the binding of a bait RBP, an RNA-based biochemical purification approach has the further advantages of eliminating additional background generated by proteins associating with the bait protein, as well as the perturbations to the cell introduced by its (ectopic) expression.

In this study, we envisioned to design a widely applicable, convenient RNA affinity tag system for the purification of *in situ*-formed RNPs that is sensitive, yet highly specific. To this end, we devised a tandem tag system consisting of an aptamer, as well as an exposed oligonucleotide. In the application of the system, tagged RNPs would first be pre-purified from crude cell lysate with high efficiency via the aptamer and then subjected to an additional purification step based on the discriminative power of antisense hybridization. Finally, to maximize robustness, applicability and efficiency of this bifunctional tag system, we rationally designed a scaffold that secures exposition of the selected structural determinants.

5.2 Results

5.2.1 Aptamer selection

For an initial pre-purification step, we surveyed the literature for specific RNA aptamers that had been repeatedly shown to be amenable to direct affinity purification through a ligand-coated matrix. The resulting shortlist of aptamers includes the dextran B512/Sephadex aptamer D8, the tobramycin aptamer J6f1, the streptavidin-aptamer S1, and the streptomycin aptamer stII (Table 5.1 and Figure 5.1 A to D), and was further evaluated according to the following selection criteria: (a) In order to prevent the final tag from becoming excessively long, the length of the aptamer sequence should preferably be below 50 nt. (b) To guarantee high efficiency of the aptamer-based pre-purification step, the dissociation constant (K_d) of aptamer and ligand should be 10^{-7} M or lower. (c) The elution should be gentle, so as not to interfere with the second purification step or downstream applications, and, in order to minimize the co-elution of unspecifically bound proteins and RNAs, ideally occur in a competitive manner.

Table 5.1 Selected RNA aptamers. Names, lengths, ligands, dissociation constants (K_d), elution methods and references to the original description of aptamers that have previously been used for the purification of RNPs are indicated. For the D8 and S1 aptamers, the indicated lengths refer to those of the minimal motifs and the full lengths (in brackets), respectively. See 7.1.4 in the appendix for the corresponding sequences.

Aptamer	Length (nt)	Ligand	K_d (nM)	Elution	Original description
D8	84 (40)	Dextran B512	250	Dextran B512	Srisawat et al., 2001; Srisawat and Engelke, 2002
J6f1	40	Tobramycin	5	Tobramycin	Hamasaki et al., 1998
S1	84 (44)	Streptavidin	70	D-Biotin	Srisawat and Engelke, 2001
stII	46	Streptomycin	1000	Streptomycin	Bachler et al., 1999

While the D8 (Figure 5.1 A) and S1 aptamers both have a length of 84 nt in their original form, functional "minimal motifs" of less than 50 nt have been described for each (Srisawat and Engelke, 2001, Srisawat and Engelke, 2002), so that all aptamers fulfill the length requirements. With a K_d in the nanomolar range (5 nM), the J6f1 (Figure 5.1 B) aptamer promises the highest affinity for its ligand, followed by the S1 aptamer (70 nM; Figure 5.1 C). The other two aptamers, D8 (250 nM) and stII (1000 nM; Figure 5.1 D), do not fulfill our binding affinity requirements. While all aptamers are amenable to native elution methods, only the S1 aptamer is not released from the matrix by competition with free (aptamer) ligand; instead, it is displaced upon the addition of streptavidin's natural ligand *D*-biotin.

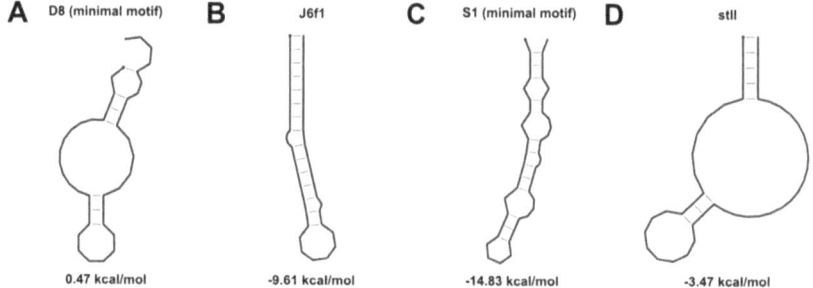

Figure 5.1 Secondary structures of RNA aptamers. The predicted centroid secondary structures of the following RNA aptamers are schematically depicted: (A) D8 (minimal motif; Srisawat and Engelke, 2002), (B) Jf61 (Hamasaki *et al.*, 1998), (C) S1 (minimal motif; Srisawat and Engelke, 2001), and (D) stII (Bachler *et al.*, 1999). Predictions are based on the reported aptamer sequences (see 7.1.4) and were made with RNAfold (Gruber *et al.*, 2008). Calculated free energies for the depicted structures are indicated. Dots indicate the 5'-termini.

Despite the unfavorable K_d between aptamer and ligand compared to the J6f1 aptamer, we chose to include the S1 aptamer for our tandem tag system, because with -14.83 kcal/mol, the predicted centroid secondary structure (the structure that is closest to the whole thermodynamic ensemble) of the S1 aptamer exhibits the lowest calculated free energy out of all shortlisted aptamers (Figure 5.1), suggesting a strong characteristic fold. This should improve the universal applicability of the tag system as it is expected that its fold, and thus its function, will not be impeded by the presence of the bait RNA. Moreover, the affinity of streptavidin to its natural ligand *D*-biotin ($k_D \sim 10^{-14}$; Green, 1990) is among the strongest non-covalent interactions known, and thus should allow captured RNPs to be eluted with utmost efficiency and specificity, minimal amounts of competitor, and without a chance to rebind the matrix. Finally, a variety of streptavidin-coated matrices are readily available from various commercial suppliers.

5.2.2 Oligonucleotide selection

Short nucleotides are able to bind RNA molecules with great discriminating power in a sequence-specific manner. For a second, high specificity purification step, we therefore decided to exploit the antisense base-pairing characteristics of oligonucleotides. Biotinylated 2'-O-methylated (2'-O-Me) RNA nucleotides have already been successfully employed for the affinity purification of RNPs (see 2.3.2.2), but the general applicability of such approaches is hampered by the immense amounts of cellular material required for the analysis of RNP composition. We reasoned, however, that the reduction in sample complexity achieved by the aptamer-mediated pre-purification step might greatly enhance the efficacy of such an approach. In order to maximize the efficiency and specificity of a hybridization-based purification strategy, the targeted sequence in the bait RNA should fulfill the following conditions: (a) The sequence should be free of strong internal secondary structures. (b) Extensive similarity or complementarity to other nucleotides present in human cells should be avoided.

For our tandem tag system to be of universal use, we set out to design such an RNA sequence and incorporate it into the tag itself – instead of scanning potential bait RNAs for regions that might fulfill these criteria. For this purpose we have used RNA Designer, a program that designs RNA sequences that fold into specific secondary structures based on user-submitted parameters (Andronescu *et al.*, 2004). Using only unpaired bases and no sequence constraints as input, we obtained five small (15 nt) and medium length (25 nt) oligonucleotides with calculated minimum free energies (MFE) of 0 kcal/mol (Table 5.2). We then performed BLAST (Altschul *et al.*, 1990) searches against the set of human reference RNAs for each of the sequences, disregarding the sequences with the lowest E values, and thus the highest chances of cross-hybridization (15v1, 15v2, 25v1, and 25v5). For each of the

Table 5.2 **Unstructured RNA sequences.** As predicted by RNA Designer (Andronescu *et al.*, 2004). Name, GC content (%), MFE, the free energy of the thermodynamic ensemble, the probability of the MFE structure in the ensemble, and the lowest E value of a Megablast search against human reference RNA sequences (Altschul *et al.*, 1990) are indicated for each of five 25- and 15-mer RNA sequences. Refer to 7.1.4 in the appendix for the corresponding sequences. # MegaBLAST search against human reference RNA sequences (accessed: October 13th, 2011).

Name	Length (nt)	GC content (%)	Minimum free energy (MFE) (kcal/mol)	Probability of MFE structure in ensemble (%)	Lowest E value[#]
15v1	15	40.0	0.00	86.3	0.76
15v2	15	46.7	0.00	97.2	0.76
15v3	15	53.3	0.00	35.5	12
15v4	15	66.7	0.00	66.3	12
15v5	15	53.3	0.00	71.4	12
25v1	25	44.0	0.00	43.0	0.24
25v2	25	52.0	0.00	13.4	3.8
25v3	25	60.0	0.00	20.5	3.8
25v4	25	52.0	0.00	53.2	3.8
25v5	25	48.0	0.00	51.3	0.96

remaining sequences, we then considered the calculated probability of the MFE structure in the thermodynamic ensemble (RNA Designer), which should give a rough indication of the degrees of (structural) freedom of the set of likely secondary structures. We reasoned that sequences with higher structural dynamics (i.e. lower probabilities of the MFE) might be more inclined to commit to binding in order to stabilize their configuration. Out of the sequences with the lowest probabilities of the MFE in the thermodynamic ensemble, 15v3 for the 15-mers (35.5%) and 25v2 for the 25-mers (13.4%), we decided to use the latter, longer sequence for initial "proof-of-principle" experiments. To specifically capture the 25v2 oligonucleotide by hybridization, we synthesized an antisense 5'-aminohexyl-3'-Cy3-linked DNA/2'-O-Me RNA hybrid oligonucleotide (25v2-as; Figure 5.2 A; see 7.1.4). The amino linker is supposed to mediate crosslinking to carboxylated microbeads via the formation of peptide bonds (Figure 5.2 B). The DNA spacer was designed to allow specific elution of captured, tagged RNAs via cleavage with *Eco*RI or an unspecific DNase. The 2'-O-Me-substituted RNA residues were included to increase stability of the oligonucleotide and are fully complementary to the 25v2 oligonucleotide in the affinity tag. Finally, incorporation of the Cy3 fluorophore should allow monitoring of bead coupling and DNase-mediated 25v2-as cleavage.

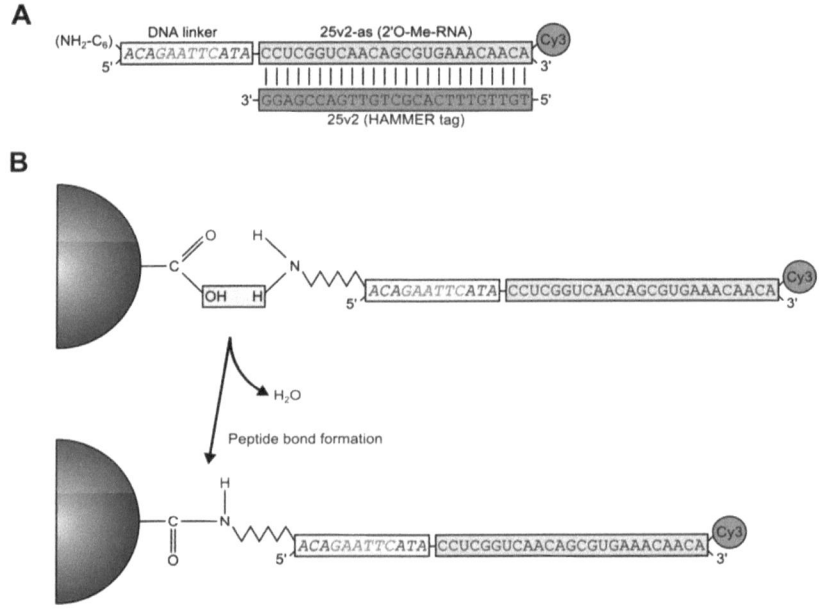

Figure 5.2 Antisense oligonucleotide 25v2-as and coupling to microbeads. (A) Schematic representation of the 25v2-as oligonucleotide used for the antisense-based purification of tagged transcripts. 25v2-as contains a 5'-terminal aminohexyl group (NH_2-C_6), a single-stranded DNA linker region (yellow box) with an *Eco*RI recognition site (red, italicized), a 25-mer 2'-O-Me RNA sequence (light red box) complementary to the 25v2 region in the tag, and a 3'-terminal Cy3 fluorophore (green circle). The 25v2 region of the tag that is complementary to 25v2-as is indicated (green box). (B) Coupling of 25vs2-as antisense oligonucleotides to carboxyl microbeads (grey). Dehydration (blue box) leads to the formation of a peptide bond between the amino group (NH_2) and the carboxyl group (COOH), covalently linking the oligonucleotide to the beads.

5.2.3 Arrangement of the HAMMER tandem affinity tag system

In order for the tandem tag system to be widely and universally applicable, both the aptamer and the oligonucleotide should always fold into the same secondary structure, regardless of the nature of the bait RNA and the relative insertion position of the tag. Furthermore, it should be ensured that the structural determinants of the tag are exposed in order for them to interact with their matrix-coated counterparts. We therefore aimed to design a scaffolding for the presentation of the structural determinants that is (a) flexible enough not to impede with their folding dynamics, and (b) strong and rigid enough to protect them from misfolding due to interaction with the bait RNA (Iioka *et al.*, 2011). For the final arrangement of aptamer and oligonucleotide, we have therefore introduced a stem structure, consisting of a stretch of 25 guanine residues opposing two stretches of cytosine residues (10 and 15 nt), on whose end is situated the S1 minimal motif, while the 25v2 oligonucleotide extends from a side-stem 15 nt into the main stem structure. Figure 5.3 A depicts the predicted centroid secondary structure. We chose to rely on G-C base pairs for the stem structure, owing to their favorable stacking interaction and hydrogen bonding characteristics compared to A-T base pairs in DNA (Sponer *et al.*, 2002; Yakovchuk *et al.*, 2006). Furthermore, we reasoned that the use of a continuous arrangement of guanines on one and cytosines on the other strand instead of alternating base pairs should prevent misalignment by allowing folding of the stem structure even if structural constraints resulting from the folding of the bait RNA require a shift of the base pairing in the stem. The final tag layout, containing the aptamer and oligonucleotide determinants as well as the stem structure, is referred to as the 'HAMMER tandem affinity tag system' or 'HAMMER tag'. Predicted base-pairing probabilities of nucleotides are high across the whole structure, particularly within the stem (>0.5), suggesting that the tandem tag will likely fold into a strong characteristic structure (Figure 5.3 B). By embedding the S1 aptamer and 25v2 oligonucleotide between restriction sites, we have designed the tag system in a modular way, thus allowing the convenient exchange of the structural determinants while keeping their general position and orientation with respect to each other and the stem (Figure 5.3 C).

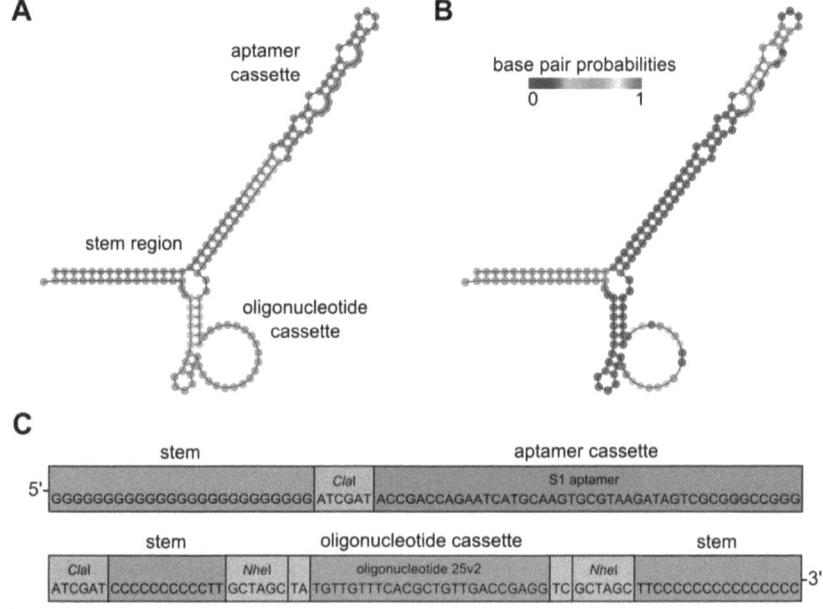

Figure 5.3 The HAMMER tandem affinity tag. (A and B) Centroid secondary structure of the HAMMER tandem affinity tag as predicted by RNAfold (Gruber *et al.*, 2008). Dots indicate the 5'-terminus. The stem region (orange), the S1 aptamer cassette (blue), and the oligonucleotide cassette (oligonucleotide: 25v2; green) are highlighted (A). Nucleotides are color-coded according to base-pair probabilities (RNAfold; B). (C) Schematic representation of the HAMMER tandem affinity tag, showing the GC-rich stem region (orange), the S1 aptamer cassette (blue), and the oligonucleotide cassette (oligonucleotide: 25v2; green), together with the corresponding DNA sequence in the sense direction. Restriction sites that can be used to exchange the S1 and oligonucleotide cassettes (*Cla*I and *Nhe*I, respectively) are indicated.

5.2.4 Purification strategy

The HAMMER tag is designed for use in a highly versatile, specific and universal tandem purification procedure which should allow the identification of protein and RNA constituents of RNPs forming on tagged RNAs by mass spectrometry and sequencing methods, respectively. A generalized schematic representation of the tandem purification process is outlined in Figure 5.4. For the chosen aptamer and oligonucleotide, the envisioned procedure includes four steps that independently introduce specificity: 1. Pre-purification of crude cell extracts through the binding of S1 aptamer to matrix-coupled streptavidin.

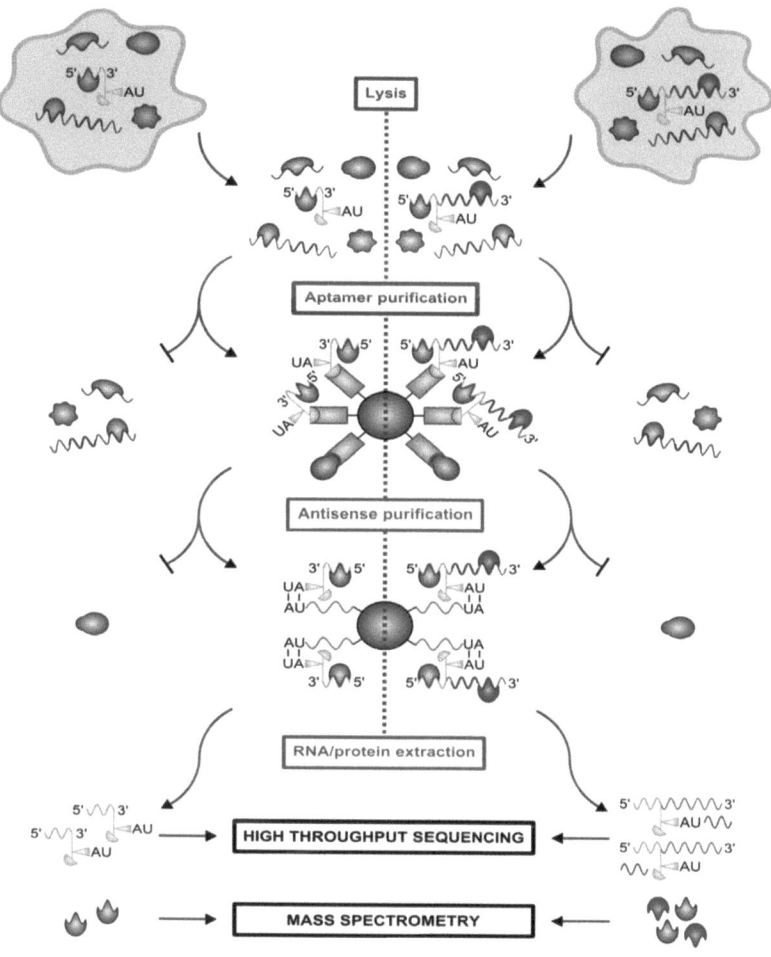

Figure 5.4 Purification of HAMMER-tagged transcripts. Schematic representation of the generalized, envisioned two-step process of RNP purification via the HAMMER tandem affinity tag. Lysates of cells expressing transcripts encoding enhanced green fluorescent protein (eGFP; green) and HAMMER (yellow), either with (right) or without (control; left) an RNA species of interest (red), are first purified via binding of the aptamer to microbeads coated with ligand (brown). HAMMER-tagged RNPs, including the RNAs and proteins that specifically associate with the RNA of interest (blue) are captured, together with all molecules that bind to eGFP, HAMMER, ligand or microbeads (grey). The majority of proteins and RNPs (grey) do not bind the ligand and are washed away. In a next step, the eluate is further purified via microbeads coupled to oligonucleotides (magenta) that are fully complementary to an exposed region in the HAMMER tag. This second step reduces the amount of unspecifically bound molecules and eliminates those proteins and RNPs that were specifically bound by the ligand in the previous step. The final eluate should be strongly enriched in HAMMER-tagged RNPs, the components of which can then be identified by (quantitative) mass spectrometry and high-throughput sequencing methods. Comparison with the control should allow the distinction of RNAs and proteins that specifically associate with the RNA species of interest from those that interact specifically or unspecifically with eGFP, the HAMMER tag, or the microbeads.

2. Competitive elution by displacement of S1 aptamer by *D*-biotin. 3. Hybridization between 25v2 and matrix-coupled 25v2-as oligonucleotides. 4. Elution by DNase-mediated cleavage of 25v2-as. Through the generation of careful control reagents, it should be possible to establish each of these steps independently.

5.2.5 Plasmid generation

In order to establish the purification procedure, we first generated a set of plasmid constructs for the expression of HAMMER-tagged RNAs (Figure 5.5 A to D). The sequence encoding the HAMMER tag was cloned into the pcDNA-5-based mammalian expression vector pTO-HA-Strep-GW-FRT (see 7.7) under the control of a tetracyclin-inducible CMV promoter (pTO-HAMMER; Figure 5.5 A). To facilitate the introduction of bait RNA sequences upstream, downstream or encompassing HAMMER, we have incorporated multiple cloning sites flanking the tandem tag. For the generation of cell lines stably expressing HAMMER-tagged RNAs, the plasmid further contains a Flippase recognition target (FRT) site, allowing convenient recombinase-mediated single site genomic integration into a range of commercially available cell lines, or – after the introduction of an FRT site into a transcriptionally active locus – a cell line of choice. An inducible promoter was chosen to allow the precise timing of the expression of HAMMER-tagged RNAs, thus minimizing potential adversary effects on transfected cells. Furthermore, it should ensure that the tag system could be used even when studying RNAs that give rise to gene products that may be toxic to cells. This is especially useful for cell lines stably expressing HAMMER-tagged RNAs.

To enable us to conveniently check for the expression of the HAMMER tag construct, we generated a variant of pTO-HAMMER in which we introduced the coding sequence of eGFP upstream of the HAMMER tag (pTO-eGFP-HAMMER; Figure 5.5 B). Several proteins and RNAs have been shown to regulate the expression of *CDKN1B* post-transcriptionally, by binding regions in its 3'-UTR (see 2.4.2). For a first "proof of principle", we therefore subcloned the *CDKN1B* 3'-UTR into the multiple cloning site downstream of the HAMMER tag (pTO-eGFP-HAMMER-*CDKN1B*-3'-UTR; Figure 5.5 C). Finally, we subcloned the eGFP-HAMMER cassette from pTO-eGFP-HAMMER into pBlueScript SK+, allowing us to generate *in vitro* transcripts of eGFP or eGFP-HAMMER from its T7 promoter (PBS-SK+-eGFP-HAMMER; Figure 5.5 D). With the help of these constructs, it should be possible to

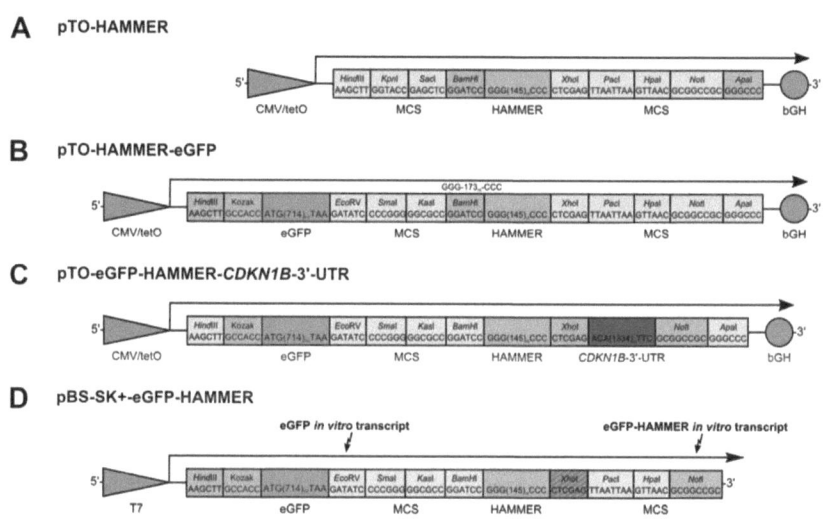

Figure 5.5 HAMMER plasmid constructs. The following plasmids containing the HAMMER tandem affinity tag have been generated: (A) pTO-HAMMER, (B) pTO-eGFP-HAMMER, (C) pTO-eGFP-HAMMER-*CDKN1B*-3'-UTR, and (D) pBS-SK+-eGFP-HAMMER. The positions and orientations of relevant DNA elements with respect to the HAMMER tandem affinity tag (blue) are indicated. The plasmids based on pTO-HA-Strep-GW-FRT (A to C; see 7.7) contain a chimeric tet operon/CMV minimal promoter (CMV/tetO; orange), allowing the tetracyclin-inducible expression of HAMMER-containing transcripts in mammalian cells (arrows). Transcription is terminated by the bovine growth hormone (bGH) terminator (A to C; orange). The T7 promoter in the plasmid based on pBlueScript-SK+ (D; Stratagene, discontinued) allows generation of eGFP and eGFP-HAMMER *in vitro* transcripts after digestion with *Eco*RV, and *Not*I respectively. The eGFP expression cassette including the Kozak sequence (B to D; dark and light green, respectively), the *CDKN1B* 3'-UTR (C; dark red), and multiple cloning sites (MCS; A to D) are indicated. DNA sequences in the sense direction are given for the restriction sites (yellow) and the eGFP Kozak sequence. Restriction sites that have been used to insert the HAMMER tandem affinity tag (A), eGFP (B), the *CDKN1B* 3'-UTR (C), and eGFP-HAMMER (D; from B) are highlighted in light red. The *Xho*I restriction site is not unique in (D; yellow and red stripes). To improve readability, overlapping restriction sites were omitted, and only one restriction enzyme was indicated for restriction sites that are recognized by more than enzyme.

establish several parameters of the purification protocol without having to cope with the problems arising from the use of complex crude cell lysates for initial tests.

5.2.6 Secondary structures of HAMMER-tagged RNAs are largely unaffected

To estimate the impact of fused RNAs on the folding of the HAMMER tag, and *vice versa*, we predicted the centroid secondary structures of the HAMMER-containing transcripts encoded on pTO-eGFP-HAMMER and pTO-eGFP-HAMMER-*CDKN1B*-3'-UTR, and compared them to the corresponding sequences without the HAMMER tag (Figure 5.6 A to D). More than 80% of the predicted eGFP secondary structure (Figure 5.6 A) are not affected by the introduction of the HAMMER tag (Figure 5.6 B); only a small region without longer stretches of nucleotides with high base-pairing probabilities and in the immediate vicinity of the tag exhibits a changed fold. As expected, the HAMMER tag folds into its characteristic shape. Furthermore, it considerably reduces the free energy of the overall structure (-263.8 kcal/mol and -196.3 kcal/mol for the tagged and untagged variants, respectively).

While the predicted structural changes resulting from introducing the HAMMER tag between the coding region and the 3'-UTR of a eGFP-*CDKN1B*-3'UTR transcript are stronger than for the eGFP-only transcript, they are also found mainly in the vicinity of the location of tag insertion and predominantly affect secondary structure elements without a high density of nucleotides with high base-pairing probabilities (Figure 5.6 C and D). A large fraction of characteristic secondary structure motifs remains unaffected. As with the eGFP transcript, introduction of the HAMMER tag decreases the free energy of the centroid structure (-416.7 kcal/mol and -391.4 kcal/mol for the tagged and untagged variants, respectively), while the structure of the tag itself is not altered.

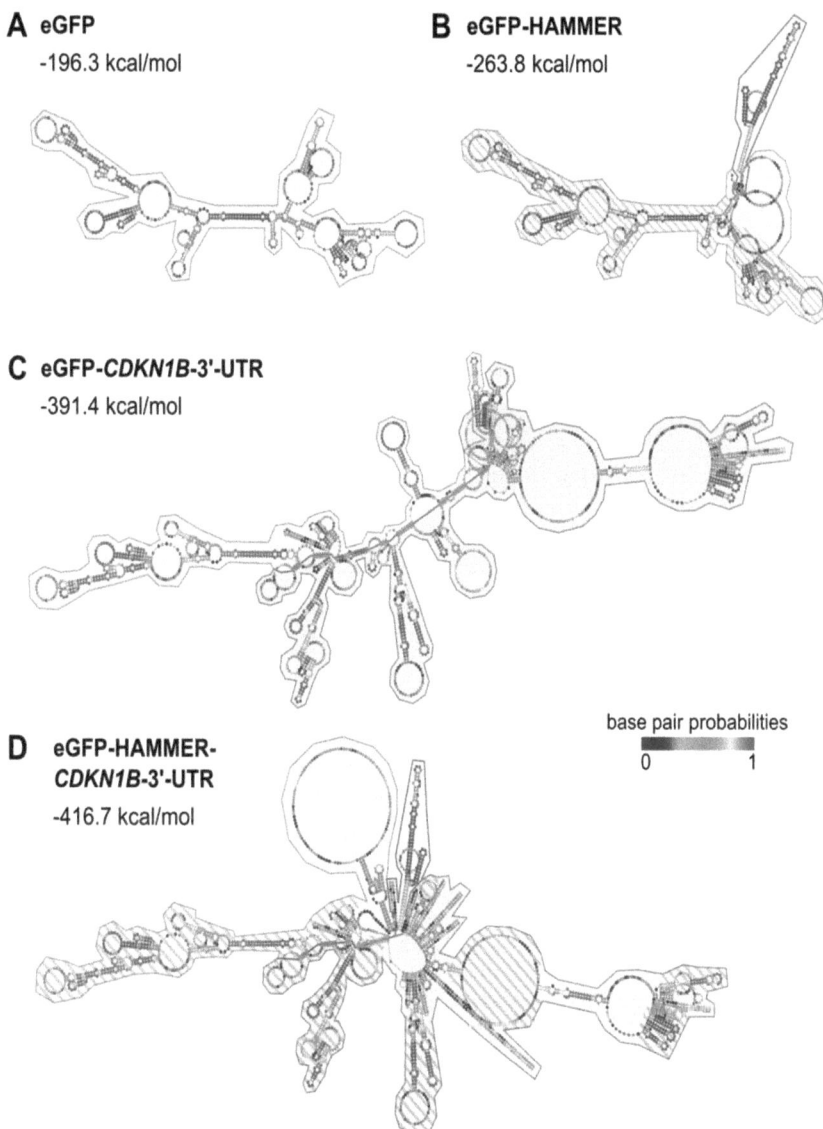

Figure 5.6 Impact of HAMMER insertion on predicted secondary structures. Centroid secondary structures of transcripts eGFP (A), eGFP-HAMMER (B), eGFP-*CDKN1B*-3'-UTR (C), and eGFP-HAMMER-*CDKN1B*-3'-UTR (D), as predicted by RNAfold (Gruber *et al.*, 2008). Calculated free energies for the depicted structures are indicated. Nucleotides are color-coded according to base-pair probabilities (RNAfold). Sequences encoding the HAMMER tandem affinity tag (yellow), the eGFP coding sequence (green) and the *CDKN1B* 3'-UTR (magenta) are highlighted. (B and D) Diagonal stripes denote regions in which the predicted secondary structures (from A and C, respectively) are not affected by incorporation of HAMMER.

When considering the predicted MFE structures instead of the centroid structures, the HAMMER-induced changes are even less pronounced. This is exemplified by a fragment of the *CDKN1B* 3'-UTR that contains several experimentally verified recognition elements for miRNAs and RNA-binding proteins (le Sage *et al.*, 2007; Galardi *et al.*, 2007; Kedde *et al.*, 2007; Kedde *et al.*, 2010). Magnification of the corresponding regions from Figure 5.6 C and D reveals that the structural context of two out of five recognition elements in the centroid secondary structures is affected by the presence of HAMMER (Figure 5.7 A and B, respectively), probably due to their close proximity to the site of the tag insertion (approximately 200-300 nt). In contrast, introduction of the tag has virtually no influence on the predicted MFE structures (Figure 5.7 C and D for the untagged and tagged variants, respectively).

Taken together, these results suggest that the HAMMER tag folds into a stable, characteristic secondary structure, irrespective of the presence of long sequence stretches adjacent to it. While parts of the tagged RNAs exhibit changes in the folding of structural motifs, these are mainly confined to regions exhibiting low frequencies of nucleotides with a high probability of base pairing and thus, presumably, unfavorable energetic properties. This is supported by the observation that structural changes are mostly found in the centroid structures, which are representative structures for the whole thermodynamic ensembles, while MFE structures are not affected. Additionally, the majority of these changes are confined to regions in close vicinity to the inserted tag. Overall, it appears that the majority of characteristic secondary structure motifs in either of the tested transcripts are not affected by tag insertion, suggesting that the potential binding of proteins and RNAs to these elements is likely not impeded. Additionally, introduction of the HAMMER tag lowered the calculated free energies of predicted centroid structures by 6 and 34% (eGFP-*CDKN1B*-3'UTR and eGFP-only transcripts, respectively), and might thus even help to stabilize such motifs.

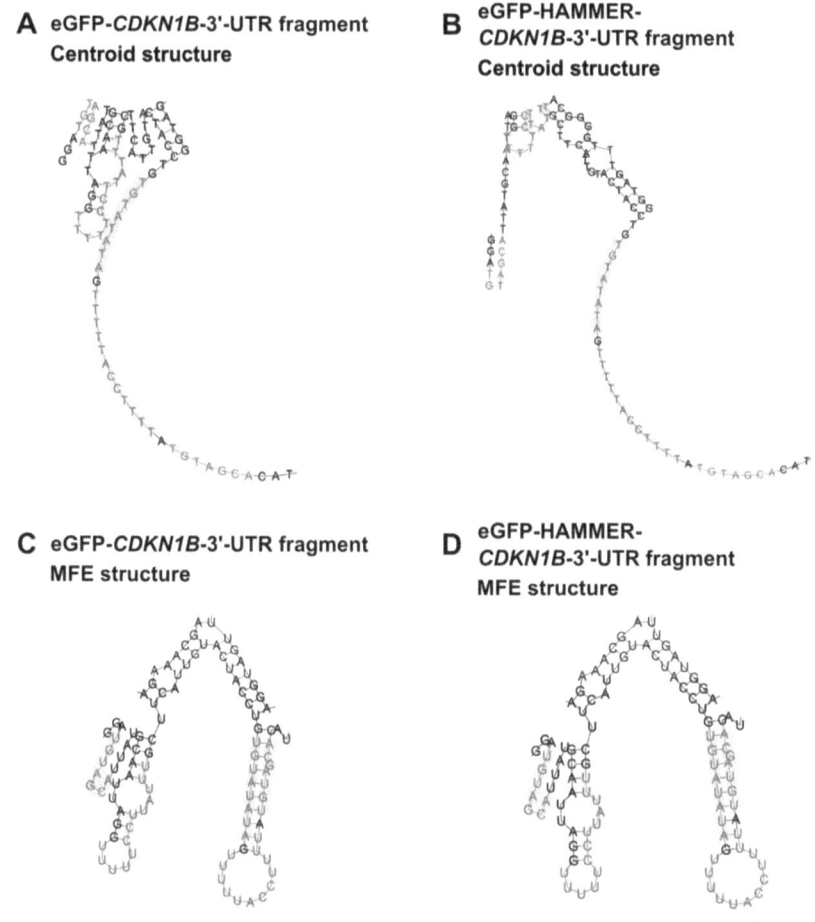

Figure 5.7 Comparison between predicted centroid and minimum free energy secondary structures. Centroid and MFE secondary structures of transcripts eGFP-*CDKN1B*-3'-UTR and eGFP-HAMMER-*CDKN1B*-3'-UTR were predicted using RNAfold (Gruber *et al.*, 2008) A fragment of the *CDKN1B* 3'-UTR containing recognition elements for miR-221/222 (green; le Sage *et al.*, 2007), Dnd1 (red; Kedde *et al.*, 2007), and Pum1 (blue; Kedde *et al.*, 2010) is depicted. (A) Centroid structure of eGFP-*CDKN1B*-3'-UTR fragment. (B) Centroid structure of eGFP-HAMMER-*CDKN1B*-3'-UTR fragment. (C) MFE structure of eGFP-*CDKN1B*-3'-UTR fragment. (D). MFE structure of eGFP-HAMMER-*CDKN1B*-3'-UTR fragment.

5.2.7 HAMMER-tagged RNAs are expressed in transiently transfected cells

In order to test whether the incorporation of HAMMER interferes with the expression of tagged RNAs, we transiently transfected Flp-In-293 cells with peGFP-HAMMER and analyzed eGFP mRNA levels by qRT-PCR, using two different primer pairs (eGFP-v2 and eGFP-v3). In both cases, transfection of peGFP-HAMMER leads to a strong and significant increase in eGFP expression compared to mock-transfected cells (Figure 5.8 A) even without induction of the CMV/tetO promoter (fold changes of $136.9 + 11.7 - 10.8$ and $100.9 + 2.3 - 2.2$ for eGFP-v2 and eGFP-v3, respectively; $P = 5.1 \times 10^{-5}$ and 2.2×10^{-7}; unpaired t-test, two-tailed). Induction with tetracyclin (2 µg/mL, 24 h) leads to a further increase in detected eGFP levels (fold changes between eGFP-HAMMER- and mock-transfected cells of $1295.1 + 171.1 - 151.2$ and $1037.3 + 119.0 - 106.8$ for eGFP-v2 and eGFP-v3, respectively; $P = 4.4 \times 10^{-9}$ and 1.6×10^{-5}; unpaired t-test, two tailed). This corresponds to fold changes between tetracyclin- and vehicle-treated cells of $9.5 + 1.3 - 1.1$ and $10.3 + 1.2 - 1.1$, for eGFP-v2 and eGFP-v3, respectively ($P = 2.2 \times 10^{-4}$ and 2.8×10^{-4}; unpaired t-test, two-tailed). The CMV/tetO promoter appears to allow residual expression of the downstream expression cassette even in the absence of tetracyclin, as the detected transcript levels account for ~10% of those measured in tetracyclin-treated samples. Furthermore, apparent residual expression of eGFP in mock-transfected cells suggests that both eGFP primer pairs are not entirely specific.

Consistently, fluorescence microscopy (Figure 5.8 B) revealed that considerable fractions of Flp-In-293 cells transiently transfected with peGFP-HAMMER or peGFP-HAMMER-*CDKN1B*-3'-UTR, but not those that did not receive plasmid (mock), were brightly green upon tetracyclin treatment (2 µg/mL; 24 h; approximately 40%, 25%, and 0%, respectively; estimated). Fractions of green fluorescent peGFP-HAMMER- and peGFP-HAMMER-*CDKN1B*-3'-UTR-transfected cells were considerably reduced upon treatment with vehicle (3% and 1.5%, respectively; estimated). Interestingly, cells that do express eGFP in the absence of tetracyclin exhibit ostensibly similar fluorescence intensities as tetracyclin-treated cells, suggesting that "promoter leakiness" may be restricted to a small subset of cells.

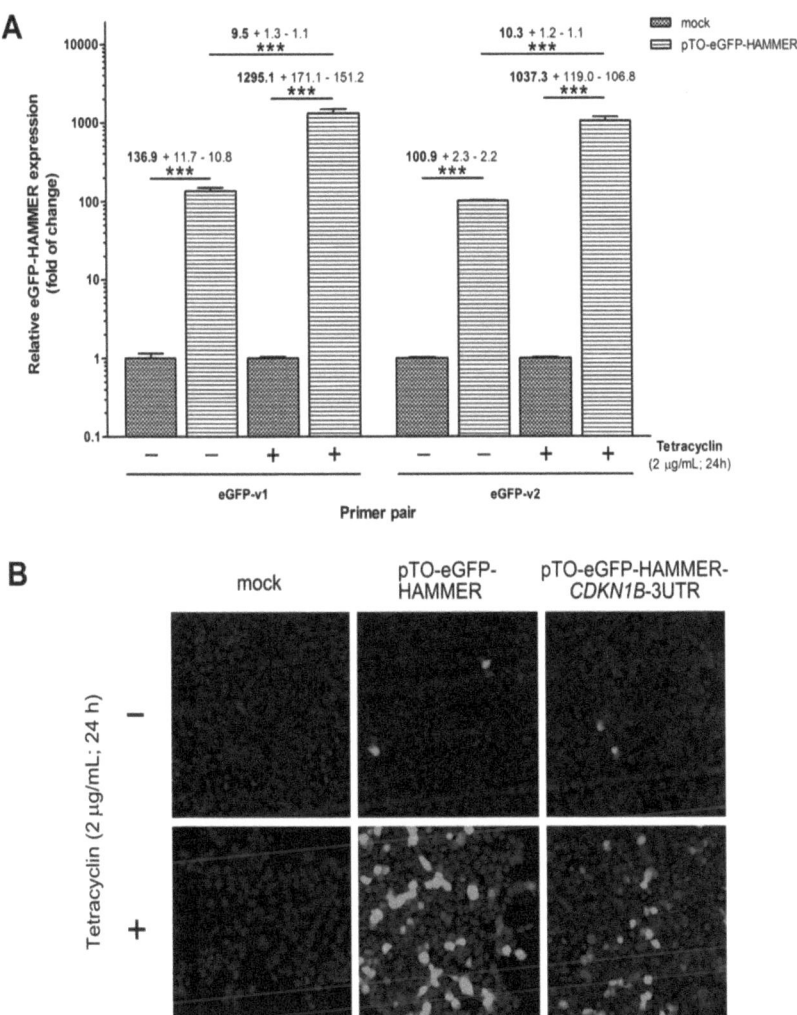

Figure 5.8 Expression of HAMMER transcripts in Flp-In-93 cells. Flp-In-293 cells were transfected with pTO-eGFP-HAMMER, pTO-eGFP-HAMMER-*CDKN1B*-3'-UTR (A only), or mock-transfected. The medium was supplemented with ethanol (EtOH; −) or tetracyclin (2 µg/mL; +) and cells were analyzed for eGFP expression 24 hours after transfection. (A) qRT-PCR experiments were performed with two different eGFP primer pairs (eGFP-v2 and eGFP-v3) and the resulting C_T values were normalized to those of ACTB. EtOH- or tetracyclin-treated pTO-eGFP-HAMMER-transfected cells were further normalized to the corresponding mock-transfected cells. Fold changes in eGFP expression ± S.D. are indicated. Experiments were performed in triplicate. Two-tailed, unpaired *t*-tests were used to calculate P values (triple asterisks denote P values <0.001). (B) Fluorescence microscopy images showing eGFP-expressing cells (green). Nuclei (blue) were stained with Hoechst 33342 dye.

Taken together, the results indicate that HAMMER-tagged transcripts are readily expressed in Flp-In-293 cells. However, since the eGFP coding region is located upstream of HAMMER in peGFP-HAMMER and peGFP-HAMMER-*CDKN1B*-3'-UTR (Figure 5.5 B and C), it cannot be concluded that expression extends across the HAMMER RNA tandem tag itself.

5.2.8 Purification of HAMMER-tagged *in vitro* transcripts via hybridization to antisense oligonucleotides

To test whether HAMMER-tagged RNA can be purified by hybridization to an antisense oligonucleotide probe, we first coupled 25v2-as oligonucleotides with carboxylated polystyrene microbeads. We then incubated this matrix with eGFP-HAMMER *in vitro* transcript with covalently bound Cy5 in binding buffer containing either 100 mM, 250 mM or no NaCl. To assess the binding of the *in vitro* transcript to the microbeads, we recorded the absorption of supernatants at $\lambda = 650$ nm (A650; Cy5) and 260 nm (A260; RNA concentration) at 0 min ("input"; t_0), after incubation for 10 min at 30°C (t_{10}), and after an additional 2 h at 4°C (t_{130}; Figure 5.9 A and B).

The measured A650 values were reduced for both time points and all salt concentrations when compared to the input (approximately 18%, 36%, and 40% decrease at t10, and 35%, 76%, and 80% at t_{130}, for 0 mM, 100 mM and 250 mM NaCl, respectively). Consistently, similar decreases were recorded for the A260 measurements (approximately 15%, 42%, and 53% decrease at t10, and 59%, 85%, and 95% at t_{130}, for 0 mM, 100 mM and 250 mM NaCl, respectively). These data indicate that 25v2-as-coupled beads are able to bind eGFP-HAMMER *in vitro* transcript. The variation in binding efficiency when using buffers with different NaCl concentrations suggests that binding occurs in a salt-dependent manner.

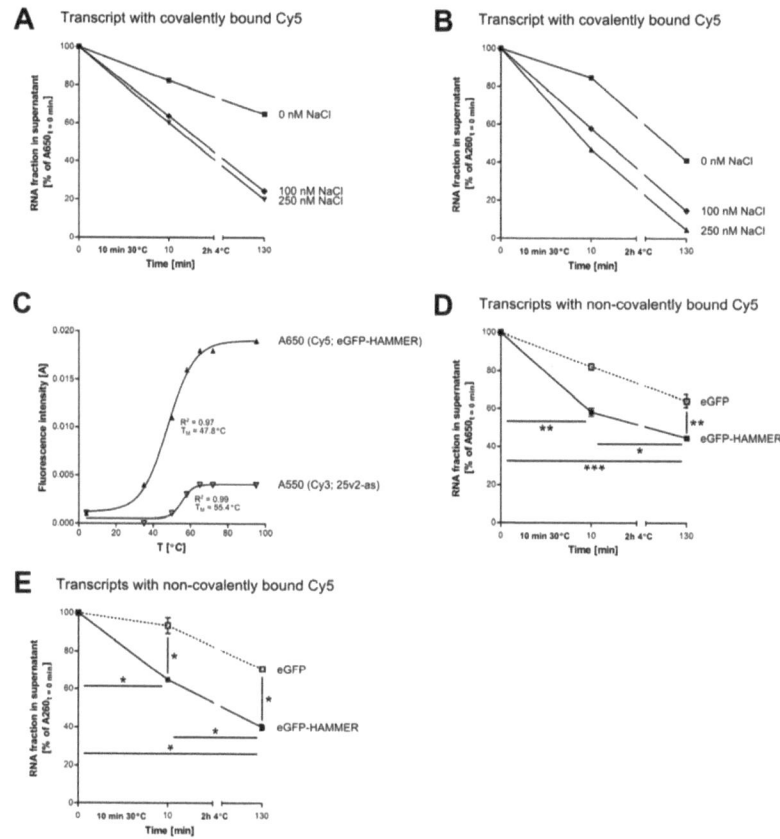

Figure 5.9 Antisense oligonucleotide purification of HAMMER *in vitro* transcripts. (A and B) *In vitro* transcripts of eGFP-HAMMER were enzymatically labeled with Cy5 and incubated with microbeads coupled to Cy3-labeled oligonucleotide 25v2-as in binding buffer containing different concentrations of sodium chloride (NaCl). Binding of transcripts to microbeads was assessed by determining free RNA levels in the supernatants (relative to input; in %) after different time points, by measuring either residual fluorescence (absorption at λ = 650 nm; A) or RNA concentration directly (absorption at λ = 260 nm; B). (C) Microbeads with bound eGFP-HAMMER *in vitro* transcripts from (A, B; 250 mM NaCl) were incubated at increasing temperatures. Release of Cy5 and Cy3 was assessed by measuring the absorption at λ = 650 nm, and 550 nm respectively. For each data set a fitted sigmoidal curve is plotted. The corresponding melting temperature (T_M) and sum of squares (R^2) are indicated. (D and E) As in (A) and (B), respectively, but data are from non-enzymatically labeled eGFP and eGFP-HAMMER *in vitro* transcripts. Two-tailed, unpaired *t*-tests were used to calculate P values (single, double and triple asterisks denote P values <0.05, 0.01, and <0.001 respectively). For clarity, asterisks were omitted for significantly reduced eGFP levels between time points t_{130} and t_{10} (*), and between time points t_{130} and t_0 (**) in (E; refer to main text).

Next, we tested whether the bound RNA could be released from the beads. Previously, we had established that the 25v2-as oligonucleotide could not be cleaved from the beads by incubation with either *Eco*RI or DNase I (Felix Schnarwiler, Institute of Pharmaceutical

Sciences, ETH Zurich; data not shown). Instead, we tested whether we could disassociate RNA:2'-O-Me RNA hybrids by heat. Aliquots of beads bound to eGFP-HAMMER transcripts were incubated for 5 min at different temperatures (T = 4, 35, 50, 58, 65, 72, and 95°C) in buffer containing 250 mM NaCl. Releases of eGFP-HAMMER transcript and 25v2-as oligonucleotide were assessed spectrophotometrically by measuring the absorption of supernatants at λ = 650 nm (A650; Cy5), and 260 nm (A550; Cy3) respectively (Figure 5.9 C). Both the A650 and A550 values increased together with the temperature, indicating that the fluorophore-containing molecules are released into the supernatant. Recorded values strictly followed sigmoidal curves with increasing temperature (R^2 = 0.97 and 0.99, for A650 and A550, respectively). The temperature at which 50% of the maximum release was achieved (melting temperature; T_M) was lower for Cy5 (47.8°C) than for Cy3 (55.4°C). The results suggest that bound RNA can be efficiently released from antisense oligonucleotide-coated beads by heat, even when using relatively moderate temperatures and short treatment time.

By comparing the binding efficiencies of eGFP transcripts with and without the HAMMER tag, we then tested whether binding of HAMMER-tagged RNA is specific. The corresponding *in vitro* transcripts were produced (Figure 5.5 D), labeled with non-covalently bound Cy5, and incubated with 25v2-as-coupled microbeads (Figure 5.2 D) in binding buffer containing 100 mM NaCl. Binding was assessed by measuring A650 and A260 after incubation for 10 min at 30°C (t_{10}), and an additional 2 h at 4°C (t_{130}; Figure 5.9 D and E). For the eGFP-HAMMER transcript, Cy5 levels in the supernatant (A650) dropped by approximately 42% between t_0 and t_{10} and 13% between t_{10} and t_{130}, adding up to a total decrease of 55% between t_0 and t_{130} (P = 0.0079, 0.0379, and 2.9 x 10^{-5} respectively; unpaired *t*-test, two-tailed). Consistently, the concentration of RNA in the supernatant (A260) decreased by approximately 35% between t_0 and t_{10} and 25% between t_{10} and t_{130}, accounting for a total drop of 60% between t_0 and t_{130} (P = 0.0297, 0.0156, and 0.0257 respectively;

unpaired t-test, two-tailed). A650 (first value) and A260 (second value) levels for the eGFP transcript also decreased, albeit to a lesser extent: 18% and 7% between t_0 and t_{10}, 18% and 22% between t_{10} and t_{130}, and a total of 36% and 29% between t_0 and t_{130} (P = 0.1373, 0.0915, 0.1499, 0.0379, 0.0882, and 0.0021 respectively; unpaired t-test, two-tailed). The decrease in Cy5 levels was around 2.3-fold (t_{10}) and 1.5-fold (t_{130}) higher for the transcript with the HAMMER tag (P = 0.0539 and 0.0098, respectively; unpaired t-test, two-tailed). With fold changes of approximately 5.3 (t_{10}) and 2.0 (t_{130}; P = 0.0256 and 0.0101, respectively; unpaired t-test, two-tailed), this difference was even more pronounced when comparing A260 measurements between eGFP-HAMMER and eGFP. The data indicate that the 25v2-as-couplex matrix is indeed able to discriminate between transcripts that contain the HAMMER sequence (eGFP-HAMMER) and those that do not (eGFP), suggesting that binding is specific. The discriminative power appears to be higher for the incubation at 30°C.

5.2.9 Purification of HAMMER-tagged *in vitro* transcripts via the S1 aptamer

To check whether HAMMER-tagged RNAs are amenable to purification with streptavidin-coated matrix via the S1 aptamer, we incubated eGFP-HAMMER or eGFP (control) *in vitro* transcripts covalently linked with Cy5 fluorophores with paramagnetic streptavidin-coated microbeads. Binding was assessed spectrophotometrically, by measuring the absorption of supernatants at λ = 650 nm (A650; Cy5) and 260 nm (A260; RNA concentration) at different time points (Figure 5.10 A and B). No decrease in the unbound RNA fractions could be detected for either transcript, neither by A650 nor by A260 measurements, indicating that the tagged RNA was not bound by streptavidin-coated beads.

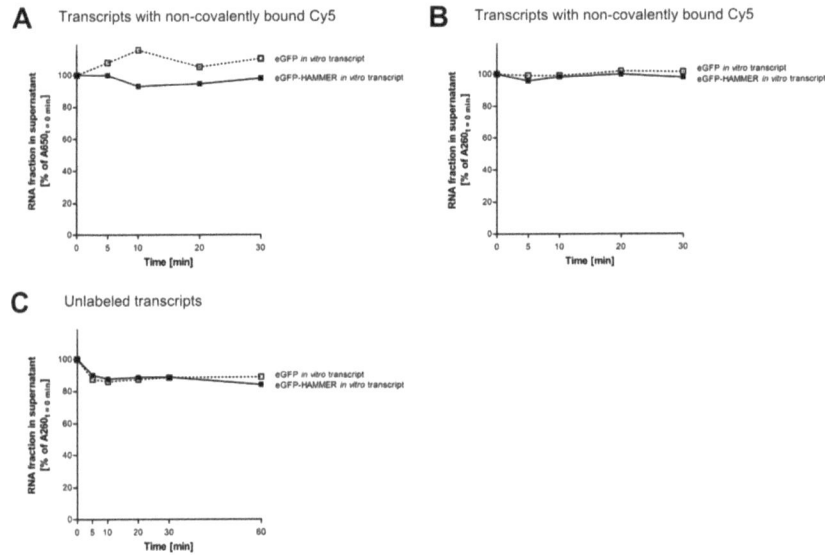

Figure 5.10 Streptavidin-S1 aptamer purification of HAMMER *in vitro* transcripts. (A to C) Cy5-labeled (A and B) and unlabeled (C) eGFP and eGFP-HAMMER *in vitro* transcripts were incubated with streptavidin-coated microbeads. Binding of transcripts to microbeads was assessed by determining free RNA levels in the supernatants (relative to input; in %) after different time points, by measuring either residual fluorescence (absorption at $\lambda = 650$ nm; A) or RNA concentration directly (absorption at $\lambda = 260$ nm; B, C).

To rule out that absence of binding is due to steric interference by aminoallyl-modified deoxyuridine triphosphates or Cy5 fluorophores, the experiment was repeated with unlabeled *in vitro* transcripts and A260 measurements only (Figure 5.10 C). Although a moderate decrease in the unbound RNA fractions in the supernatants could be observed after 5 min of incubating streptavidin-coated microbeads with transcripts (approximately 10% with respect to the input), there was no difference in A260 decrease between the eGFP-HAMMER and eGFP *in vitro* transcripts. For both transcripts, levels of unbound RNA were stable after prolonged incubation. These results indicate that HAMMER-tagged transcripts cannot be purified via the S1 aptamer under the tested reaction conditions.

5.3 Discussion

Here we present a strategy for the tandem purification of RNPs via the HAMMER tag, a novel RNA tag system consisting of a previously described RNA streptavidin aptamer and an unstructured oligonucleotide designed for antisense-based purification. The tag system was set up in a modular way, conveniently allowing the exchange of aptamer or oligonucleotide while keeping the general layout, and includes a scaffold that was rationally designed to expose and stabilize the tag components, as well as allow them to fold reliably into a characteristic secondary structure. By flanking the tag with multiple cloning sites, bait RNA sequences can be introduced on either side of it, or even around it. As all components of the tag system are based on unmodified RNA nucleotides, the tag can be expressed in cells, thus allowing RNP formation on tagged RNAs *in situ*. The envisioned two-step procedure involves pre-purification via a streptavidin-coated matrix, competitive elution with *D*-biotin, a second purification step relying on hybridization of the exposed oligonucleotide with a matrix-coated DNA/2'-O-Me RNA antisense oligonucleotide, and DNase-mediated elution, and should thus be highly specific. The gentle elution methods should enable the identification of RNP components by a variety of downstream applications, such as mass spectrometry and RNA sequencing.

5.3.1 Expression of tagged transcripts

We have verified the expression of HAMMER-tagged eGFP in a HEK293-derived cell line by qRT-PCR and fluorescence microscopy. But as the tag insertion site was located downstream of the eGFP coding in the tested plasmids, it was not clearly shown that the tag itself was properly expressed. The presence of long stretches of guanine and cytosine residues in DNA has been reported to negatively affect transcription both *in vitro* (Belotserkovskii *et al.*, 2010) and in yeast (Kim and Jinks-Robertson, 2011), and thus could pose a potential problem for the expression of the HAMMER tag with its GC-rich stem structure. However, the fluorescence microscopy and qRT-PCR results suggest normal amounts of eGFP both on the mRNA and the protein level. Furthermore, both *in vitro* transcription of HAMMER-tagged RNAs and propagation of HAMMER-containing plasmids yielded usual amounts of products of the expected sizes, thus suggesting that both RNA and DNA polymerases can process the sequences encoding the tag system. Nevertheless, the ability of cells to express intact copies of HAMMER-tagged transcripts and the HAMMER tag itself should be carefully studied in future experiments.

5.3.2 Capturing of tagged transcripts by antisense hybridization

Proper T7 RNA polymerase-mediated transcription of the tag is further supported by the observation that antisense oligonucleotide-coupled microbeads are able to specifically and efficiently bind HAMMER-tagged *in vitro* transcripts when compared to an untagged control. Under the tested incubation conditions, transcripts with Cy5-labeled aminoallyl-modified UTPs were bound efficiently in the presence of approximately physiological ionic strength, while binding of transcripts labeled with intercalated Cy5 was less efficient. Low salt concentrations considerably reduced binding whereas elevated salt levels slightly increased the efficiency of capturing tagged RNAs. Incubation at 30°C appears to have favorable binding kinetics compared to incubation in the cold. Similarly, the difference between binding efficiencies of tagged and untagged *in vitro* transcripts was higher after the 30°C incubation step (2- to 5-fold lower for the latter) compared to end point measurements (1.5- to 2-fold less for untagged transcript), suggesting that specificity is higher at the elevated temperature. These results are consistent with a report stating that annealing rates between RNA-RNA heterodimers decrease considerably at temperatures around 30 degrees below the melting temperature (Patzel and Sczakiel, 1999). Considering the determined melting temperature of the RNA:2'-O-Me RNA hybrid of approximately 50°C, incubation temperatures above or around room temperature should be preferable for the oligonucleotide hybridization procedure. Given the high efficiency of the binding and an approximately two-fold enrichment of signal over noise in the pilot experiments, the oligonucleotide purification strategy appears to be promising. While its efficacy in purifying cellular transcripts remains to be proven, the envisioned aptamer-based pre-purification step should be able to reduce sample complexity considerably so that it can be hoped that efficiency and specificity will not dramatically decrease compared with the purification of the *in vitro* transcript. Furthermore, by testing different antisense oligonucleotide pairs, binding buffer compositions and incubation/reaction conditions, it should be possible to further optimize specificity and efficiency of the procedure.

5.3.3 Elution of transcripts immobilized by hybridization

As the cleavage of bead-coupled 25v2-as oligonucleotide by either *Eco*RI or DNase I was not successful, elution of captured HAMMER-tagged Cy5-labelled *in vitro*-transcript was achieved by "melting" RNA:2'O-Me RNA hybrids at elevated temperatures. Based on the measurements of released Cy5, a melting curve was recorded which, as expected, exhibits a sigmoidal shape, strongly resembling the thermal denaturation of DNA. Unexpectedly, labeled 25v2-as was also released upon heating up the beads, possibly due to residual, uncoupled 25v2-as oligonucleotide binding to the bead surface. While thermal elution was efficient even at moderate temperatures and short incubation times, a more gentle and specific elution method would be preferable to exclude the possibility of RNA alkaline hydrolysis, particularly in the presence of divalent cations and a relatively high pH in the buffer. One explanation for the failure of endonuclease-mediated cleavage could be the inaccessibility of the DNA linker due to its limited length (12 bp) and proximity to the beads. Moreover, while it has been reported that both enzymes, as well as type II restriction enzymes in general, are able to process single-stranded DNA, cleavage usually occurs at reduced rates (Nishikagi *et al.*, 1985; Bischofberger *et al.*, 1987; Sutton *et al.*, 1997; Latham, Ambion, unpublished). Possible strategies for establishing an elution method relying on endonuclease-mediated cleavage of the DNA linker include the use of endonucleases with increased specificity for single-stranded DNA, extending the incubation time, optimization of reaction/buffer conditions, the use of a longer DNA linker, and the addition of short DNA fragments complementary to the DNA linker sequence. The use of excessive amounts of endonuclease is not recommended if mass spectrometric analysis of the eluate is intended, as signals resulting from fragments of highly abundant proteins may mask those of lesser abundant proteins, as has been reported for example for the mass spectrometric analysis of the human blood plasma proteome (Anderson and Anderson, 2002; Atkins *et al.*, 2002). Similarly, the choice of buffer components should be carefully considered so as not to interfere with downstream applications.

5.3.4 Aptamer-mediated purification of tagged transcripts

In initial experiments aimed at establishing a protocol for S1 aptamer-mediated purification, no binding of HAMMER-tagged Cy5-labeled *in vitro* transcripts to streptavidin-coated microbeads could be detected. It is conceivable that the presence of fluorophore complexes in the labeled transcripts prevents proper folding of the S1 aptamer and, consequently, binding to the streptavidin-coated matrix. Indeed, binding occurred when unlabeled HAMMER-tagged and untagged *in vitro* transcripts were used instead, albeit indiscriminately and with low efficiency. One possible explanation is that the aptamer, although properly folded, is inaccessible. However, this is not likely to be the culprit in this case, since we tested the procedure with *in vitro* transcripts, largely in the absence of factors that might potentially interfere with the folding of the tag. Moreover, the same *in vitro* transcripts appeared to be well accessible in the antisense oligonucleotide purification experiments. While we cannot rule out that the observed effects are caused by unfavorable reaction conditions (binding buffer composition or incubation times/temperature), the very low binding efficiency (<15%), the binding kinetics (plateau reached after 5 min) and the complete absence of specificity suggest a more fundamental problem: As was reported in the original description of the S1 aptamer (Srisawat and Engelke, 2001), integration into bait RNAs sometimes led to dysfunctional aptamers, even when the predicted secondary structures suggested proper folding. To rule out positional effects, the tag should first be inserted into different locations. If the problem persists, it may be necessary to include a more flexible linker between the stem structure and the aptamer. Alternatively, either the full-length S1 aptamer (Srisawat and Engelke, 2001) or a different aptamer could be employed. The tobramycin aptamer J6f1 might constitute a reasonable choice as its binding affinity to the ligand is reported to be in the nanomolar range (Hamasaki *et al.*, 1998), although both the affinity and the specificity of the aptamer have been called into question (Verhelst *et al.*, 2004).

5.3.5 Reflections on tag folding and insertion

So far we have generated plasmids that allow *in vitro* transcription of HAMMER-tagged transcripts, as well as their expression in mammalian cells. Incorporation of the tag downstream of an eGFP coding sequence as well as between an eGFP coding sequence and a *CDKN1B* 3'-UTR did cause minor changes in the predicted secondary structures of the corresponding transcripts, while the tag structure itself was not affected by the presence of the bait RNAs. Changes in the fold of bait RNAs were predominantly located in the vicinity of the location of tag insertion and mostly affected regions with a low density of nucleotides with high base-pairing probability. In order to avoid or minimize such adverse effects, it might be advisable to introduce the tag in those regions of potential bait RNAs that exhibit exposed structural motifs with low free energies. Better yet, generating a set of constructs in which one and the same bait RNA is tagged at two or more different positions should increase the chances of proper aptamer and bait RNA folding and should also reduce the chance of steric interference between the tag structure and potential binding factors of the bait RNA. Moreover, results obtained from the parallel purification of bait RNAs using tags introduced at different locations should increase both the robustness and reliability of the results. In general, it is recommended to consult secondary structure prediction software such as RNAfold (Gruber *et al.*, 2008) or MC-Fold/MC-Sym (Parisien and Major, 2008) for choosing appropriate insertion positions.

5.3.6 Limitations of RNA secondary structure prediction algorithms

While changes in the folding of an RNA in different sequence contexts can be estimated with some confidence by employing secondary structure prediction algorithms, the accuracy and relevance of a specific predicted structure cannot be assessed without empirical structural data. Regarding the absence of binding to the ligand-coated matrix, it might be possible that the aptamer requires interconversion between two or more stable folds to allow binding to the ligand in an "induced fit"-like manner (Williamson, 2000; Leulliot and Varani, 2001). It is conceivable that the presence of the stem structure or steric interference of the bait RNA traps a non-functional intermediate structure in a local minimum of the 'folding landscape' (Chen and Dill, 2000; Solomatin *et al.*, 2010), thus restricting structural dynamics. Furthermore, it should be kept in mind that classical nucleic acid secondary structure prediction algorithms are mostly based on base pairing and stacking interactions of nucleobases and may, on their own, be of very limited use for the estimation of RNA conformations in RNA-protein complexes. Therefore, a more in-depth structural analysis should be performed, making use of advanced prediction software that may take into account the tertiary structure of the aptamer (e.g. MC-Fold/MC-Sym; Parisien and Major, 2008), folding dynamics (e.g. BarMap; Hofacker *et al.*, 2010), and its interaction with protein ligands (e.g. catRAPID; Bellucci *et al.*, 2011). Ideally, it should include the experimental determination of the solution structure of the functional aptamer, which can then be used to fine-tune prediction parameters. In this regard, low (Jiang *et al.*, 1997) and high (Jiang and Patel, 1998) resolution structures of tobramycin bound to a J6f1-related aptamer (Wang and Rando, 1995; Wang *et al.*, 1996), as well as a crystal structure of the stII aptamer in complex with streptomycin (Tereshko *et al.*, 2003) already exist.

5.3.7 Conclusion

If it is possible to overcome the addressed technical difficulties and establish a working tandem purification protocol, the HAMMER tandem affinity tag system holds promise to be a versatile and widely applicable tool for the analysis of *in situ*-formed RNPs in a wide range of cell-based or even whole organism systems. In a next step, optimization and streamlining of the purification and analysis procedures may lead to improved reliability and reproducibility, as well as reduced costs and hands-on time. To this end, the use of paramagnetic beads as the matrix in both purification steps may help to achieve a certain level of automation which, together with, for example, the use of the Gateway cloning system (Invitrogen) and the construction of "UTRome libraries" may allow medium- to high-throughput analysis of RNPs. Furthermore, additional information may be obtained, e.g. by the incorporation of crosslinking methods (Niranjanakumari *et al.*, 2002), which should enable the simultaneous identification of *cis*-regulatory messages in tagged bait RNAs. Finally, the use of the tag system is not limited to RNP purification: By targeting regions of the tag with fluorescently labeled antisense oligonucleotides, it should be possible to study RNP localization by fluorescence *in situ* hybridization methods (Walker *et al.*, 2011).

5.4 Materials and Methods

5.4.1 Tag design and bioinformatics

RNA secondary structure predictions were made with RNAfold (http://rna.tbi.univie.ac.at/cgi-bin/RNAfold.cgi; Gruber *et al.*, 2008), using default settings and an energy parameter model established by Andronescu *et al.*, (Andronescu *et al.*, 2007). Unstructured RNA sequences (15-mers and 25-mers) were designed using RNA Designer (http://www.rnasoft.ca/cgi-bin/RNAsoft/RNAdesigner/rnadesign.pl; Andronescu *et al.*, 2004; structure: string of 15 or 25 times ".", sequence constraints: string of 15 or 25 times "N", temperature: 37°C, number of sequences: 5, GC content for paired/unpaired regions: 50%, random number seed: 1).

5.4.2 Plasmids

The designed HAMMER tandem affinity tag sequence (HAMMER-synth; see 7.1.4) was ordered from Mr. Gene (http://mrgene.com/). To generate pTO-HAMMER, the insert sequence of the obtained pMK-based vector was verified by sequencing and subcloned into pTO-HA-Strep-GW-FRT (kindly provided by Alexander Wepf, Institute of Molecular and Systems Biology, ETH Zurich; see 7.7) via *Bam*HI and *Apa*I, replacing its HA and StrepIII protein tags as well as the Gateway cassette with the HAMMER tandem affinity tag. The coding sequence of eGFP was amplified from pTO-HA-Strep-GW-FRT-eGFP (kindly provided by Alexander Wepf; Institute of Molecular and Systems Biology, ETH Zurich) using the primers *Hind*III-Kozak-eGFP-fwd (containing a restriction site for *Hind*III as well as a Kozak site; see 7.1.1) and eGFP-rev-MCS-*Bam*HI (containing several restriction sites, including one for BamHI; see 7.1.1). The amplicon was purified with the QIAquick PCR

Purification kit (QIAGEN, Cat. No. 28104) and inserted into pCR-Blunt II-TOPO (Invitrogen), using the Zero Blunt TOPO PCR Cloning Kit (Invitrogen, Cat. No. K2830-20) according to the manufacturer's recommendations. After verifying correct insertion by sequencing of the corresponding region, the eGFP cassette was subcloned into pTO-HAMMER via *Hind*III and *Bam*HI. The region containing eGFP and the HAMMER sequence were subcloned into pBlueScript SK+ (Stratagene, discontinued) via *Hind*III and *Not*I to generate pBS-SK+-eGFP-HAMMER. The *CDKN1B*/p27 3'-UTR (nt 1070-2403 of RefSeq mRNA NM_004064.3) was subcloned from pCR-Blunt II-TOPO-*CDKN1B*-3'-UTR (see 4.4.2) via *Xho*I and *Not*I to generate pTO-eGFP-HAMMER-*CDKN1B*-3'-UTR.

5.4.3 Cell culturing

Flp-In-293 cells (Invitrogen, Cat. No. R750-07; kindly provided by Alexander Wepf, Institute of Molecular and Systems Biology, ETH Zurich) were cultured in Dulbecco's modified Eagle's medium (DMEM; Invitrogen, Cat. No. 41966) supplemented with 10% fetal bovine serum (Invitrogen, Cat. No. 10270-106), 1x antibiotic-antimycotic (Invitrogen, Cat. No. 15240-062), and 1 mg/mL zeocin (Invitrogen, Cat. No. R250-01) in 5% CO_2 at 37°C.

5.4.4 Quantitative reverse transcription PCR

2.5 mL of a Flp-In-293 (Invitrogen, Cat. No. R750-07) cell suspension (2×10^5 cells/mL in DMEM supplemented with 10% fetal bovine serum and 1x antibiotic-antimycotic) were seeded into the wells of an 6-well plate (TPP, Switzerland, Cat. No. 92006). 16 hours later, cells were transfected with 2 µg of peGFP-HAMMER, or mock-transfected, using polyethylenimine (Polysciences, Inc., Cat. No. 23966): DNA and polyethylenimine stock solution (1 mg/mL in water) were diluted to 15 µg/mL and 60 µg/mL, respectively, in DMEM. Both solutions were incubated for 10 min at room temperature, and then mixed at equal volumes (mass ratio polyethylenimine to DNA = 4:1; final polyethylenimine concentration = 30 µg/mL). After incubation for 20 min at room temperature, transfection solutions were added to the cells. 24 hours after transfection, medium was replaced with DMEM supplemented with 10% fetal bovine serum, 1x antibiotic-antimycotic, and either 2 µg/mL tetracyclin (2 mg/mL stock solution in EtOH) or a corresponding volume of EtOH only. 24 hours after transfection, cells were washed with phosphate-buffered saline (PBS; Invitrogen, Cat. No. 14190) and RNA was isolated by using the RNeasy Mini kit (QIAGEN, Cat. No. 74106) with RNase-free DNase I (QIAGEN, Cat. No. 79254) digestion according to the manufacturer's recommendations (elution in nuclease-free water). RNA concentrations were determined spectrophotometrically with a NanoDrop 1000 (Thermo Fisher Scientific Inc.). cDNA was prepared from total RNA (50 ng per reaction) using the High Capacity cDNA Reverse Transcription Kit (Applied Biosystems, Cat. No. 4368814) according to the manufacturer's recommendations. Quantitative PCR reactions were performed in triplicate, using FastStart Universal SYBR Green Master (Rox; Roche, Cat. No. 04913914001) in an AB 7900 HT Fast Real-Time PCR System (Applied Biosystems). The following primer pairs were used at 0.1 nM: ACTB (ACTB-SYBR-fwd, ACTB-SYBR-rev), eGFP-v1 (eGFP-SYBR-v1-fwd, eGFP-SYBR-v1-rev), and eGFP-v2 (eGFP-SYBR-v2-fwd, eGFP-SYBR-v2-rev). Quantification was performed using the $2^{-\Delta\Delta CT}$ method (Schmittgen and Livak, 2008), with ACTB serving as a reference for the normalization of expression levels.

5.4.5 Fluorescence microscopy

250 µL of a Flp-In-293 (Invitrogen, Cat. No. R750-07) cell suspension (4 x 10^4 cells/mL in DMEM supplemented with 10% fetal bovine serum and 1x antibiotic-antimycotic) were seeded into the chambers of an 8-chamber slide (BD Falcon, Cat. No. 354108). 16 hours later, cells were transfected with 100 ng of peGFP-HAMMER or peGFP-HAMMER-

CDKN1B-3'-UTR, or mock-transfected, using polyethylenimine (Polysciences, Inc., Cat. No. 23966): DNA and polyethylenimine stock solution (1 mg/mL in water) were diluted to 15 µg/mL and 60 µg/mL, respectively, in DMEM. Both solutions were incubated for 10 min at room temperature, and then mixed at equal volumes (mass ratio polyethylenimine to DNA = 4:1; final polyethylenimine concentration = 30 µg/mL). After incubation for 20 min at room temperature, transfection solutions were added to the cells. 24 hours after transfection, medium was replaced with DMEM supplemented with 10% fetal bovine serum, 1x antibiotic-antimycotic, and either 2 µg/mL tetracyclin (2 mg/mL stock solution in EtOH) or a corresponding volume of EtOH only. 24 hours later, Hoechst dye 33342 (Invitrogen, Cat. No. H3570) was added to the medium, and after 10 min cells were analyzed with an Axiovert 200 M (Zeiss) fluorescence microscope. Images were acquired with the AxioVision software (Zeiss; release 4.7) and overlay images were created with Adobe Photoshop CS5.

5.4.6 *In vitro* transcription and labeling

pBS-SK+-eGFP-HAMMER was digested with either *Eco*RV or *Not*I for the generation of eGFP and eGFP-HAMMER *in vitro* transcripts, respectively. Linearized plasmid was run on a 1% TAE agarose gel and purified using the MinElute Gel Extraction Kit (QIAGEN, Cat. No. 28604) according to the manufacturer's recommendations (elution in nuclease-free water). *In vitro* transcripts with non-covalently bound Cy5 (eGFP and eGFP-HAMMER) were produced with the MEGAscript T7 Kit (Ambion, Cat. No. AM1333) according to manufacturer's recommendations, using 3.4 µg linearized plasmid in 100 µL volume as template. *In vitro* transcribed RNA was isolated by using the RNcasy Mini kit with DNase I digestion according to the manufacturer's recommendations (elution in nuclease-free water). 20 µg purified RNA were Cy5-labeled with the Turbo Labeling Kit Cy5 (Arcturus, Cat. No. KIT0610) according to the manufacturer's recommendations. *In vitro* transcript with covalently bound Cy5 (eGFP-HAMMER only) was produced from 750 ng linearized plasmid in 14 µL volume using the Amino Allyl MessageAmp II aRNA Amplification Kit with Cy5 (Ambion, AM1796) according to the manufacturer's recommendations, starting from the *in vitro* transcription step. RNA quality and the correct lengths of products were assessed by agarose gel electrophoresis. Yield and efficiency of Cy5 incorporation were determined with a NanoDrop 1000.

5.4.7 S1 aptamer purification

120 µL of streptavidin-coupled microbeads (Streptavidin MagneSphere Paramagnetic Particles, Cat. No. Z5481) were incubated three times (20 min, 850 rpm; followed by collection of beads using a magnet) in 200 µL blocking buffer (20 mM Tris-HCl, pH 8.0; 100 mM NaCl; 5% $^v/_v$ glycerol; 5 mM $MgCl_2$, pH 8.0; 0.1% Triton X-100; 0.1 % BSA; 0.1 mg/ml heparin; 0.1 mg/ml *E. coli* tRNA; 40 U/ml RNasin Plus, Promega, Cat. No. N2615), and resuspended in 50 µL of the same buffer. 10 pmol of eGFP-HAMMER *in vitro* transcript (calculated molecular weight without label ≈ 302 kilodalton; kDa) were dissolved in 50 µL of binding buffer (20 mM Tris-HCl, pH 8.0; 100 mM NaCl; 5% $^v/_v$ glycerol; 0.1% Triton X-100; 5 mM $MgCl_2$, pH 8.0;), incubated at 65°C for 5 min, and allowed to cool down slowly to RT. The RNA solution was combined with the blocked microbeads (total volume = 100 µL) and incubated for 30 or 60 min at 4°C. 5 µL aliquots were removed after 0, 5, 10, 20, 30, and 60 (where applicable) min and analyzed by spectrophotometry with a NanoDrop 1000. Absorption at λ = 260 nm and, in the case of Cy5-labeled transcripts, 650 nm was recorded in order to determine the unbound RNA fraction.

5.4.8 Oligonucleotide synthesis

5'-aminohexyl-3'-Cy3-linked DNA/2'-O-Me RNA hybrid 25v2-as (see 7.1.4) oligonucleotides were synthesized from the 3'- to the 5'-terminus with a MerMade 12 (Bioautomation Corporation; MerMade 12 v2.3.7 software) on controlled-pore glass solid support (Glen UnySupport, 500 Å, 50 µmol/g; Glen Research) under standard conditions. Each synthesis cycle involved deblocking of the dimethoxytrityl group with 5% dichloroacetic acid in dichlorethane, coupling of 0.1 M phosphoamidites in acetonitrile (SAFC Proligo) upon activation with 0.25 M 5-ethylenthiotetrazole (360 s for 2'-O-Me RNA amidites, 180 s for all other phosphoamidites), capping of uncoupled amidites with a mixture of 10% acetic anhydride in tetrahydrofuran ("Capping A"; SAFC Proligo) and 10% N-methylimidazole and 10% pyridine in water ("Capping B"; SAFC Proligo) in order to avoid synthesis of incomplete product, and stabilization of the phosphoamidite linkage by oxidation with 0.02 M iodine in a mixture (7:2:1) of tetrahydrofuran, pyridine and water (RFLC Proligo). 5'-amino-modifier C6 and Cy3 phosphoamidites were from Glen Research, and DNA and 2'-O-Me RNA phosphoamidites were from Thermo Fisher Scientific, Inc. The product was cleaved and deprotected in a screw cap vial with a mixture (1:1) of aqueous ammonia (33%; Fluka) and ethanolic methylamine (33%; Fluka) for 30 min at 65°C and washed once with a mixture (1:1) of water and EtOH. The combined supernatants were dried in a SpeedVac (Savant SPD 2010; Thermo Electronic), dissolved in 0.1 M triethylammonium acetate and purified by semi-preparative reversed phase high-performance liquid chromatography (RP-HPLC; Agilent 1200 HPLC system, Agilent Technologies) on a C18 column (XBridge OST C18 2.5 µm; 4.6 x 50 mm; Waters), based on the presence of the monomethoxytrityl group in the 5'-amino-C6, which was only available in the full-length product. Relevant fractions were dried in a SpeedVac, the residue solubilized in 40% acetic acid and incubated at RT for 60 min to remove monomethoxytrityl groups. After liquid evaporation in a SpeedVac, residue was dissolved in 0.1 M hexafluoroisopropanol with 8.6 mM triethylamine. Purity (>99%) and identity (12721.2 Da found; 12723.8 Da calculated; deviation = 0.020%) were verified by RP-HPLC on a C18 column (XBridge OST C18 2.5 µm;

2.1 x 50 mm; Waters), followed by liquid chromatography-mass spectrometry (LC-MS; Agilent 6130 Single Quad, Agilent Technologies) analysis.

5.4.9 Preparation of antisense oligonucleotide matrix

5.6 nmol of Cy3-labeled 25v2-as oligonucleotide (in 800 μL water) were coupled to 29.2 mg (1167 μL of suspension) of carboxylated 0.75-1 μm polystyrene microbeads (Polysciences, Inc., Cat. No. 07759) by using the PolyLink Protein Coupling Kit (Polysciences, Inc., Cat. No. 24350-1) according to manufacturer's recommendations (reagent volumes were scaled accordingly). Incubation was performed for two hours at RT with gentle shaking (850 rpm). The suspension was centrifuged (10 min, 1000 g, RT) at the beginning (0 min; resuspended in same buffer) and the end of the incubation time (2 h, RT, 850 rpm), and supernatants were subjected to spectrophotometric analysis with a NanoDrop 1000 ($\lambda = 550$ nm) in order to determine the fraction of unbound oligonucleotide. Coupled beads were resuspended in 1.2 mL of PolyLink Wash/Storage Buffer (2.8 nmol bound oligonucleotides per mL suspension) and stored at 4°C.

5.4.10 Purification by antisense oligonucleotide hybridization

100 μL of 25v2-as-coupled microbead suspension were equilibrated for 5 min in binding buffer (20 mM Tris-HCl, pH 8.0; 5% $^V/_V$ glycerol; 5 mM $MgCl_2$, pH 8.0; 0.1% Triton X-100) including 100 mM NaCl (unless mentioned otherwise). Beads were collected by centrifugation (5 min, 3,000 g, RT), resuspended in 100 μL of the same buffer, and mixed with 10 pmol of the indicated *in vitro* transcript. Suspensions were incubated for 10 min at 30°C, and then for 2 h at 4°C. 15 μL aliquots were removed after 0, 10, and 130 min, centrifuged (2 min, 13,000 g, RT), and analyzed by spectrophotometry with a NanoDrop 1000. Absorption at $\lambda = 260$ nm and 650 nm were recorded in order to determine the fractions of

unbound RNA. To assess the release of bound RNA by increasing temperature, microbeads with bound Cy5-labeled eGFP-HAMMER *in vitro* transcript in buffer with 250 mM NaCl were topped up with additional buffer to a volume of 128 µL and divided into eight aliquots with 15 µL each. Aliquots were incubated for 5 min at either 4, 35, 50, 58, 65, 72, or 95°C, centrifuged (2 min, 13,000 g, RT), and analyzed by spectrophotometry with a NanoDrop 1000. Absorption at $\lambda = 550$ nm (Cy3) and 650 nm (Cy5) were recorded in order to determine the fractions of unbound 25v2-as oligonucleotide and Cy5-labeled eGFP-HAMMER *in vitro* transcript, respectively.

5.5 Contributions

André P. Gerber, Alexander Kanitz, Luca Schenk and Jonathan Hall (Institute of Pharmaceutical Sciences, ETH Zurich) contributed to the design of the HAMMER RNA tandem tag. Mauro Zimmermann (Institute of Pharmaceutical Sciences, ETH Zurich), under the supervision of Jonathan Hall, prepared the 25v2-as oligonucleotides, and Felix Schnarwiler (Institute of Pharmaceutical Sciences, ETH Zurich), under the supervision of André P. Gerber and Alexander Kanitz, performed the coupling of 25v2-as oligonucleotides to microbeads. Alexander Kanitz, under the supervision of André P. Gerber and Michael Detmar, performed all other experiments. Katarzyna Hunt (Institute of Pharmaceutical Sciences, ETH Zurich), under the supervision of Alexander Kanitz, André P. Gerber and Luca Schenk, helped with the generation of plasmids, the expression analysis, as well as aptamer and antisense oligonucleotide purification experiments. Michael Detmar (Institute of Pharmaceutical Sciences, ETH Zurich) provided general support and, together with André P. Gerber, supervised Alexander Kanitz and Luca Schenk. Alexander Wepf, supervised by Matthias Gstaiger (Institute of Molecular and Systems Biology, ETH Zurich), contributed plasmids and cell lines as indicated in the main text.

6 Concluding Remarks

Since the dawn of molecular biology half a century ago, the exploration of transcriptional and post-translational control of gene expression has absorbed the vast majority of the available capacities for the study of gene regulation. It does therefore, perhaps, not come as a surprise that the discovery of the first miRNA in 1993 (Lee *et al.*, 1993) was neglected as a worm idiosyncrasy by the scientific community. It required the advent of genomic techniques and the insights gained from the sequencing of the first higher organisms to alert scientists of the previously disregarded, almost completely undiscovered 'parallel universe' of post-transcriptional gene regulation. And even today, and despite the increased interest in post-transcriptional gene regulation that has been largely caused by the 're-discovery' and subsequent appreciation of the far-reaching implications of microRNA (miRNA) regulation, the full significance of post-transcriptional gene regulatory networks (GRNs) may still be underestimated. Nevertheless, lessons learned from studying transcriptional and post-translational networks have accelerated the rate of progress in the field so that now, for example, several elegant and sophisticated methods exist that allow targets and recognition elements of RNA-binding proteins and miRNAs to be identified on a transcriptome-wide scale. However, such "top down" approaches are limited to spotting 'multiple output' network motifs only. In contrast, gene-centered "bottom up" strategies aimed at unveiling 'multiple input' motifs, i.e. the identification of multiple *trans*-acting factors regulating a single (m)RNA, are less straight-forward and often rely on chance findings or predictive tools. Yet considering that thousands of laboratories around the world largely rely on the study of individual genes to identify and develop leads for potential drug targets, the demand for detailed insights into the post-transcriptional regulatory impact exerted on specific messages is high.

Conclusion

In this work, we used a biased, integrative approach relying on bioinformatics predictions and empirical data to identify *cis*-regulatory elements and *trans*-acting factors that repress the expression of a well-established human angiogenesis factor with high medical relevance. Importantly, the levels of all repressors were significantly reduced in a particular tumor type that expresses high levels of the targeted angiogenesis factor, suggesting a potential role in disease. Our findings thus validate the feasibility of the outlined procedure for the identification of combinatorial control motifs. Furthermore, as the used data and prediction algorithms are readily available, the strategy could be useful for many laboratories studying physiological or pathological aspects of a limited number of genes. However, the biased, time-intensive and error-prone nature of the procedure severely hampers its scalability and limits its use to 'hypothesis-driven' research.

In a second study, we therefore laid the foundation for a widely applicable method for the unbiased discovery of proteins and RNAs associated with a transcript of interest. The method relies on a modular RNA tandem affinity tag that we rationally designed to facilitate its appendage to RNAs of interest while maintaining a high stability, folding characteristics and exposure of the affinity determinants. The tag system is nucleic acid-based and thus able to be expressed inside cell lines of interest, ensuring the unimpeded assembly of near-native ribonucleoprotein complexes which can then be purified by immobilization on ligand-coated matrices in a two-step process, guaranteeing a high level of specificity. Gentle elution methods should permit the analysis of co-purified proteins and RNAs by transcriptomics and proteomics approaches. While we were not able to provide a 'proof of principle', we believe that our tag design holds great promise for the unbiased identification of 'multiple input' motifs of post-transcriptional GRNs, as the availability of 'discovery-driven' complementary top-down and bottom-up approaches is indispensable for the comprehensive characterization of the basic motifs of such networks.

By integrating the data generated by the application of these and other methods into the available network models, they can be gradually expanded to include other, more complex motifs, such as feedback or feed-forward loops. The ultimate goal, however, would be the reconciliation of all different types of gene regulatory networks into a single, robust composite network of gene regulation. Such a model would have broad implications for medical and pharmaceutical sciences, synthetic biology, computer sciences and technology, and even the study of social networks in economics. Moreover, it could help us to approach an answer to perhaps the oldest question in the biological sciences: How can chemical systems sustain the remarkable organizational complexity that we call life?

7 Appendix

7.1 Nucleotide sequences

7.1.1 Nucleotides used for cloning

PCR primer sequences used for the cloning of the indicated (final) plasmids are listed. Used restriction or attB Gateway recombination sites are underlined. Nucleobases that are part of a coding sequence or untranslated region are italicized.

psiCHECK-2-CDKN1B-3'-UTR (and mutated version), peGFP-HAMMER-*CDKN1B*-3'UTR

CDKN1B-3'-UTR-fwd	5'-<u>CTCGAGACAG</u> *CTCGAATTAA GAATATGTTT CC*-3'
CDKN1B-3'-UTR-rev	5'-<u>GCGGCCGCGA</u> *AGTTTTCTTT ATTGATTACT TAATGTG*-3'

psiCHECK-2-VEGFA-3'-UTR (and mutated versions)

VEGFA-3'-UTR-fwd	5'-TCA<u>CTCGAGG</u> *TCCCGGCGAA GAGAAGAG*-3'
VEGFA-3'-UTR-rev	5'-CAT<u>GCGGCCG</u> *CTCAATGGAG AAGGAGAAACCA*-3'

peGFP-HAMMER

*Hind*III-Kozak-eGFP-fwd	5'-GATT<u>AAGCTT</u> GCCACCATGG *TGAGCAAGGG CGAGG*-3'
eGFP-rev-MCS-*Bam*HI	5'-<u>GGATCCGGCG</u> CCCCCGGGGA TATCTTACTT *GTACAGCTCG TCCATGC*-3'

pTO-HA-Strep-GW-FRT-Pum2

Pum2-CDS-attB-fwd	5'-GGGGACAAGT <u>TTGTACAAAA AAGCAGGCTT</u> *AATGAATCAT GATTTTCAAG C*-3'
Pum2-CDS-attB-rev	5'-GGGGACCACT <u>TTGTACAAGA AAGCTGGGTT</u> *ACAGCATTCC ATTTGG*-3'

7.1.2 Nucleotides used for mutagenesis

Oligonucleotide sequences used for the mutagenesis of the VEGFA 3'-UTR are listed. Mutated residues are underlined. Deleted residues are marked by hyphens. Note that the PRE mutations were introduced sequentially, starting with PRE3, and followed by PRE2 and PRE1.

VEGFA-PRE TM

VEGFA-PRE3-MUT-fwd	5'-CAACTTGTAT TTGTGTGTAT ATATATATAT ATATGTTTAA CAATATATGT GATTCTGATA AAATAGACAT TGCTATTCTG-3'
VEGFA-PRE3-MUT-rev	5'-CAGAATAGCA ATGTCTATTT TATCAGAATC ACATATAT<u>C</u> TTAAACATAT ATATATAT ATACACACAA ATACAAGTTG-3'
VEGFA-PRE2/3-MUT-fwd	5'-TCTACATACT ATATATATAT TTGGCAACTT GTATTTGTGA CAATATATAT ATATATATGT TTAA<u>C</u>AATAT ATGTGATTCT G-3'
VEGFA-PRE2/3-MUT-rev	5'-CAGAATCACA TATATT<u>GTTA</u> AACATATATA TATATATATT GT<u>C</u>ACAAATA CAAGTTGCCA AATATATATA TAGTATGTAG A-3'
VEGFA-PRE1-MUT-fwd	5'-CTCTTGCTCT CTTATTTGTA CCGGTTTTA<u>C</u> AATATAAAAT

	TCATGTTTCC AATCTCTCT-3'
VEGFA-PRE1-MUT-rev	5'-AGAGAGATTG GAAACATGAA TTTTATAT<u>TG</u> <u>T</u>AAAACCGGT ACAAATAAGA GAGCAAGAG-3'

VEGFA-MRE-MUT

VEGFA-MRE-MUT-fwd	5'-GTGTGTATAT ATATATATAT ATGTTTATGT ATATATGTGA T---GATAAA ATAGACATTG CTATTCTGTT TTTTATATGT AAAAACAAA-3'
VEGFA-MRE-MUT-rev	5'-TTTGTTTTTA CATATAAAAA ACAGAATAGC AATGTCTATTT TATC---ATC ACATATATAC ATAAACATAT ATATATATAT ATACACAC-3'

7.1.3 Nucleotides used for quantitative reverse transcription PCR (SYBR Green)

Oligonucleotides used for SYBR Green-based qRT-PCR are listed.

ACTB

ACTB-SYBR-fwd	5'-TCACCGAGCG CGGCT-3'
ACTB-SYBR-rev	5'-TAATGTCACG CACGATTTCC-3'

eGFP-v1

eGFP-SYBR-v1-fwd	5'-CCTGAAGTTC ATCTGCACCA-3'
eGFP-SYBR-v1-rev	5'-GAAGAAGTCG TGCTGCTTCA-3'

eGFP-v2

eGFP-SYBR-v2-fwd	5'-CGACGGCAAC TACAAGAC-3'
eGFP-SYBR-v2-rev	5'-TAGTTGTACT CCAGCTTGTG C-3'

7.1.4 Nucleotides for the HAMMER RNA tandem tag

RNA aptamer sequences. The used aptamer (S1 minimal motif) is marked with an asterisk.

D8 (minimal motif)	5'-GUCCGAGUAA UUUACGUUUU GAUACGGUUG CGGAACUUGC-3'
J6f1	5'-ACCGACCAGA AUCAUGCAAG UGCGUAAGAU AGUCGCGGGC CGGG-3'
S1 (minimal motif)*	5'-GGCUUAGUAU AGCGAGGUUU AGCUACACUC GUGCUGAGCC-3'
stII	5'-GGAUCGCAUU UGGACUUCUG CCCAGGGUGG CACCACGGUC GGAUCC-3'

Unstructured RNA oligonucleotides. Generated using RNA Designer (http://www.rnasoft.ca/cgi-bin/RNAsoft/RNAdesigner/rnadesign.pl; Andronescu *et al.*, 2004). The used sequence (25v2) is marked with an asterisk.

15v1	5'-UUUAUCUUCA GCUGG-3'
15v2	5'-UAUUGUGUCC CUCUC-3'
15v3	5'-UUGCCCGUAG GAUCA-3'
15v4	5'-AUGCGCCGCU CGAGA-3'
15v5	5'-UGAGUCACGU CCGUA-3'

Appendix

```
25v1        5'-UUUAUCUUCA CUUGACUAGC CGGCU-3'
25v2*       5'-UGUUGUUUCA CGCUGUUGAC CGAGG-3'
25v3        5'-CAGACCCUAG GAUUACGUGC ACCGG-3'
25v4        5'-AUGCGCCGCU CGAGAAACAC AAUUG-3'
25v5        5'-UGAGUCACGU CCGUAAACCU AAUGC-3'
```

HAMMER sequence. Obtained from Mr. Gene (http://mrgene.com/).

```
HAMMER-synth    5'-GGATCCGGGG GGGGGGGGGG GGGGGGGGGG GATCGATACC
                   GACCAGAATC ATGCAAGTGC GTAAGATAGT CGCGGGCCGG
                   GATCGATCCC CCCCCCCTTG CTAGCTATGT TGTTTCACGC
                   TGTTGACCGA GGTCGCTAGC TTCCCCCCCC CCCCCCCCTC
                   GAGTTAATTA AGTTAACGCG GCCGCGGGCC C-3'
```

Sequence of the 5'-amino-3'-Cy3 DNA/2'-O-methylated RNA hybrid for antisense oligonucleotide purification. 2'O-methylated RNA nucleotides are complementary to 25v2 and are underlined. Synthesized by Mauro Zimmermann (Institute of Pharmaceutical Sciences, ETH Zurich).

```
25v2-as     NH2-(C6)-5'-ACAGAATTCA TACCUCGGUC AACAGCGUGA AACAACA-3'-Cy3
```

7.2 MicroRNA mimics and antisense inhibitors

Ordering information and, where applicable, identifiers and sequences of the mature miRNAs are indicated.

Name	MicroRNA ID	Sequence	Supplier	Product ID	Pre-/Anti-miR ID
Cy3 dye-labeled Pre-miR Negative Control # 1	n/a	n/a	Ambion	AM17120	n/a
Cy3 dye-labeled Anti-miR Negative Control # 1	n/a	n/a	Ambion	AM17011	n/a
Pre-miR Negative Control #1	n/a	n/a	Ambion	AM17110	n/a
Anti-miR Negative Control #1	n/a	n/a	Ambion	AM17010	n/a
Pre-miR-361-5p	hsa-miR-361-5p	UUAUCAGAAUC UCCAGGGGUAC	Ambion	AM17100	PM10085
Anti-miR-361-5p	hsa-miR-361-5p	UUAUCAGAAUC UCCAGGGGUAC	Ambion	AM17000	AM10085

7.3 Commercial quantitative reverse transcription PCR assays

In addition to the ordering information, an identifier of the targeted gene (Entrez), and, where available, the binding regions of the primers and probes as well as the amplicon lengths (in nucleotides) are indicated.

Name	System	Entrez	Supplier	Product ID	Assay ID	Binding	Length
ACTB	TaqMan	60	Applied Biosystems	4326315E	n/a	exon 1	171
CHM	TaqMan	1121	Applied Biosystems	4351372	Hs01114163_m1	exons 9-10	102
miR-126	TaqMan	406913	Applied Biosystems	4427975	002228	n/a	n/a
miR-205	TaqMan	40698	Applied Biosystems	4427975	000509	n/a	n/a
miR-20b	TaqMan	574032	Applied Biosystems	4427975	001014	n/a	n/a
miR-34a	TaqMan	407040	Applied Biosystems	4427975	000426	n/a	n/a
miR-361-5p	TaqMan	494323	Applied Biosystems	4427975	000554	n/a	n/a

Appendix

Name	System	Entrez	Supplier	Product ID	Assay ID	Binding	Length
miR-93	TaqMan	407050	Applied Biosystems	4427975	001090	n/a	n/a
ACTB	SYBR	60	QIAGEN	QT01680476	n/a	n/a	104
PUM1	SYBR	9698	QIAGEN	QT0002941	n/a	exons 5-6	73
PUM2	SYBR	23369	QIAGEN	QT00067760	n/a	exons 3/4	89
RNU6B	TaqMan	26826	Applied Biosystems	4427975	001093	n/a	n/a
VEGFA 3'-terminus	TaqMan	7422	Applied Biosystems	4331182	Hs03929054_s1	exon 8	134
VEGFA exon 3	TaqMan	7422	Applied Biosystems	4331182	Hs99999070_m1	exon 3	63

7.4 MicroRNAs predicted to target VEGFA

MicroRNAs predicted to have recognition elements in the VEGFA 3'-UTR are listed. Target predictions were from the following web services: microRNA.org (MR; Betel *et al.*, 2010), TargetScan (TS; Friedman *et al.*, 2009), DIANA-microT (µT; Maragkakis *et al.*, 2009), miRDB (DB; Wang, 2008), and MicroCosm (MC; Griffiths-Jones *et al.*, 2008). For each MRE, the genome coordinates (chromosome 6, + strand), the 3'-UTR region as defined in the main text (Reg.; 1 = conserved region 1, NC = non-conserved region, 2 = conserved region 2), the relative position to the start site of the 3'-UTR (Pos.; based on GenBank RefSeq entry NM_001025366.2), the particular algorithms and the total number of algorithms predicting the MRE (#; first number), as well as the total number of algorithms for which target predictions for the corresponding miRNA were available in the accessed information (#; second number), are indicated. For MREs previously subjected to experimental analysis, the corresponding references are indicated.

MicroRNA	Genome coordinates	Reg.	Pos.	MR	TS	µT	DB	MC	#	Reference
hsa-miR-1976	43752300-43752312	1	1	yes	n/d	n/d		n/d	1/2	
hsa-miR-125a-5p	43752303-43752331	1	4	yes					1/5	
hsa-miR-1321	43752309-43752326	1	10	yes		n/d		n/d	1/3	
hsa-miR-125b	43752310-43752331	1	11	yes					1/5	
hsa-miR-4319	43752315-43752331	1	16	yes	n/d	n/d	n/d	n/d	1/1	
hsa-miR-218-1*	43752323-43752343	1	24	yes	n/d	n/d		n/d	1/2	
hsa-miR-4299	43752340-43752356	1	41	yes	n/d	n/d	n/d	n/d	1/1	
hsa-miR-136*	43752359-43752381	1	60	yes	n/d	n/d		n/d	1/2	
hsa-miR-29b-2*	43752368-43752391	1	69	yes	n/d	n/d		n/d	1/2	
hsa-miR-34a	43752371-43752398‡	1	72						0/5	Ye et al., 2008
hsa-miR-140-5p	43752373-43752392‡	1	74						0/5	Ye et al., 2008
hsa-miR-885-3p	43752377-43752398	1	78	yes	yes				2/5	
hsa-miR-34b*	43752377-43752399‡	1	78						0/5	Ye et al., 2008
hsa-miR-26b*	43752407-43752428	1	108	yes	n/d	n/d		n/d	1/2	
hsa-miR-3169	43752411-43752432	1	112	yes	n/d	n/d	n/d	n/d	1/1	
hsa-miR-499-5p	43752411-43752434	1	112	yes					1/5	
hsa-miR-1299	43752420-43752441	1	121	yes	yes	n/d		n/d	2/3	
hsa-miR-1246	43752422-43752440	1	123	yes		n/d		n/d	1/3	
hsa-miR-205*	43752426-43752451	1	127	yes	n/d	n/d	n/d	n/d	1/1	
hsa-miR-516b	43752435-43752441*	1	136	yes					1/5	
hsa-miR-205	43752437-43752456	1	138	yes	yes†	yes	yes	yes	5/5	Ye et al., 2008; Wu et al., 2009
hsa-miR-1236	43752441-43752461	1	142	yes		n/d	yes	n/d	3/3	
hsa-miR-877*	43752442-43752462	1	143	yes	n/d	n/d		n/d	1/2	
hsa-miR-579	43752447-43752463*	1	148		yes†				1/5	
hsa-let-7i*	43752449-43752474	1	150	yes	n/d	n/d		n/d	1/2	
hsa-miR-4279	43752450-43752465	1	151	yes	n/d	n/d	n/d	n/d	1/1	
hsa-miR-520b	43752453-43752481$	1	154		yes†	yes			2/5	
hsa-miR-520c-3p	43752453-43752481$	1	154		yes†	yes			2/5	
hsa-miR-520d-3p	43752453-43752481$	1	154		yes†	yes			2/5	
hsa-miR-302c	43752456-43752482	1	157	yes	yes†				2/5	
hsa-miR-372	43752457-43752482	1	158	yes	yes†	yes			3/5	Ye et al., 2008
hsa-miR-33b*	43752459-43752481	1	160	yes	n/d	n/d	n/d		1/2	
hsa-miR-520a-3p	43752460-43752482	1	161	yes	yes†	yes			3/5	
hsa-miR-302a	43752460-43752482	1	161	yes	yes†				2/5	
hsa-miR-302b	43752460-43752482	1	161	yes	yes†				2/5	
hsa-miR-593	43752460-43752480	1	161	yes					1/5	
hsa-miR-93	43752461-43752483	1	162	yes	yes†				2/5	Ye et al., 2008
hsa-miR-520e	43752462-43752482	1	163	yes	yes†	yes			3/5	
hsa-miR-373	43752462-43752482	1	163	yes	yes†				2/5	Ye et al., 2008
hsa-miR-520g	43752462-43752484	1	163	yes	yes†				2/5	Ye et al., 2008
hsa-miR-519d	43752463-43752483	1	164	yes	yes†	yes		yes	4/5	
hsa-miR-106a	43752463-43752483	1	164	yes	yes†	yes			3/5	Hua et al., 2006; Ye et al., 2008
hsa-miR-20a	43752463-43752483	1	164	yes	yes†	yes			3/5	Hua et al., 2006; Ye et al., 2008

Appendix

MicroRNA	Genome coordinates	Reg.	Pos.	MR	TS	µT	DB	MC	#	Reference
hsa-miR-17	43752463-43752483	1	164	yes	yes†				2/5	Hua et al., 2006; Ye et al., 2008; Lei et al., 2009
hsa-miR-20b	43752463-43752483	1	164	yes	yes†				2/5	Hua et al., 2006; Ye et al., 2008
hsa-miR-526b*	43752463-43752483	1	164	yes	n/d	n/d	n/d		1/2	
hsa-miR-520h	43752464-43752484	1	165	yes	yes†				2/5	Ye et al., 2008
hsa-miR-106b	43752465-43752483	1	166	yes	yes†	yes			3/5	Hua et al., 2006; Ye et al., 2008
hsa-miR-302e	43752466-43752482	1	167	yes	yes†	n/d		n/d	2/3	
hsa-miR-711	43752468-43752489	1	169	yes	n/d	n/d	n/d	n/d	1/1	
hsa-miR-186*	43752468-43752486	1	169	yes	n/d	n/d	n/d		1/2	
hsa-miR-302d	43752475-43752481*	1	176		yes†				1/5	Ye et al., 2008
hsa-miR-613	43752489-43752517$	1	190			yes			1/5	
hsa-miR-1249	43752491-43752497*	1	192		yes	n/d		n/d	1/3	
hsa-miR-516a-3p	43752507-43752513*	1	208		yes				1/5	
hsa-miR-492	43752518-43752542	1	219	yes					1/5	
hsa-miR-3126-5p	43752525-43752545	1	226	yes	n/d	n/d	n/d	n/d	1/1	
hsa-miR-520a-5p	43752528-43752550	1	229	yes					1/5	
hsa-miR-525-5p	43752529-43752550	1	230	yes					1/5	
hsa-miR-652	43752544-43752564	1	245	yes					1/5	
hsa-miR-3163	43752548-43752569	1	249	yes	n/d	n/d	n/d	n/d	1/1	
hsa-miR-922	43752556-43752580	1	257	yes					1/5	
hsa-miR-31	43752558-43752579	1	259	yes					1/5	
hsa-miR-15a	43752559-43752582	1	260	yes	yes†	yes			3/5	
hsa-miR-103	43752559-43752581	1	260	yes		yes			2/5	
hsa-miR-107	43752559-43752581	1	260	yes		yes			2/5	
hsa-miR-545	43752559-43752580	1	260	yes					1/5	
hsa-miR-16	43752560-43752582	1	261	yes	yes†	yes			3/5	Karaa et al., 2009
hsa-miR-503	43752560-43752582	1	261	yes	yes†				2/5	
hsa-miR-15b	43752561-43752582	1	262	yes	yes†	yes			3/5	
hsa-miR-424	43752561-43752582	1	262	yes	yes†	yes			3/5	
hsa-miR-497	43752562-43752582	1	263	yes	yes†				2/5	
hsa-miR-195	43752563-43752582	1	264	yes	yes†	yes		yes	4/5	
hsa-miR-374b*	43752563-43752584	1	264	yes	n/d	n/d	n/d		1/2	
hsa-miR-646	43752564-43752582	1	265	yes	yes†				2/5	
hsa-miR-423-3p	43752572-43752594	1	273	yes	yes				2/5	
hsa-miR-1180	43752579-43752600	1	280	yes		n/d		n/d	1/3	
hsa-miR-141*	43752581-43752603	1	282	yes	n/d	n/d	n/d		1/2	
hsa-miR-548e	43752593-43752614	1	294	yes	yes	n/d		n/d	2/3	
hsa-miR-548a-3p	43752594-43752614	1	295	yes	yes				2/5	
hsa-miR-548f	43752596-43752614	1	297	yes	yes	n/d		n/d	2/3	
hsa-miR-323b-5p	43752618-43752640	1	319	yes	n/d	n/d	n/d	n/d	1/1	
hsa-miR-1293	43752619-43752641	1	320	yes	yes	n/d		n/d	2/3	
hsa-miR-363*	43752624-43752645	1	325	yes	n/d	n/d	n/d		1/2	
hsa-miR-139-5p	43752627-43752633*	1	328		yes				1/5	
hsa-miR-299-3p	43752628-43752649	1	329	yes	yes†		yes		3/5	Jafarifar et al., 2011
hsa-miR-567	43752629-43752651	1	330	yes	yes				2/5	Jafarifar et al., 2011
hsa-miR-934	43752629-43752635*	1	330		yes				1/5	
hsa-miR-3149	43752635-43752657	1	336	yes	n/d	n/d	n/d	n/d	1/1	
hsa-miR-609	43752635-43752641*	1	336		yes				1/5	Jafarifar et al., 2011
hsa-miR-297	43752636-43752656	1	337	yes	yes		yes		3/5	Jafarifar et al., 2011
hsa-miR-3171	43752636-43752658	1	337	yes	n/d	n/d	n/d	n/d	1/1	
hsa-miR-340	43752641-43752662	1	342	yes					1/5	
hsa-miR-410	43752643-43752663	1	344	yes	yes				2/5	
hsa-miR-374a	43752654-43752677	1	355	yes	yes		yes		3/5	
hsa-miR-889	43752655-43752675	1	356	yes					1/5	
hsa-miR-374b	43752656-43752677	1	357	yes	yes		yes		3/5	
hsa-miR-369-3p	43752657-43752676	1	358	yes	yes				2/5	
hsa-miR-410	43752658-43752678	1	359	yes	yes				2/5	
hsa-miR-3145	43752669-43752697	1	370	yes	n/d	n/d	n/d	n/d	1/1	
hsa-miR-590-3p	43752670-43752691	1	371	yes					1/5	
hsa-miR-539	43752676-43752699	1	377	yes					1/5	
hsa-miR-1270	43752679-43752701	1	380	yes		n/d		n/d	1/3	
hsa-miR-202	43752680-43752699	1	381	yes					1/5	
hsa-miR-135a*	43752681-43752702	1	382	yes	n/d	n/d	n/d		1/2	
hsa-miR-1303	43752681-43752702	1	382	yes	n/d			n/d	1/3	
hsa-miR-3121	43752682-43752704	1	383	yes	n/d	n/d	n/d	n/d	1/1	
hsa-miR-620	43752682-43752701	1	383	yes					1/5	
hsa-miR-376c	43752684-43752704	1	385	yes					1/5	
hsa-miR-3163	43752686-43752707	1	387	yes	n/d	n/d	n/d	n/d	1/1	
hsa-miR-142-5p	43752687-43752707	1	388	yes					1/5	
hsa-miR-340	43752687-43752708	1	388	yes					1/5	
hsa-miR-410	43752689-43752709	1	390	yes	yes				2/5	
hsa-miR-577	43752689-43752709	1	390	yes					1/5	
hsa-miR-1259	43752690-43752710	1	391	yes		n/d		n/d	1/3	
hsa-miR-568	43752692-43752710	1	393	yes					1/5	
hsa-miR-3145	43752698-43752721	1	399	yes	n/d	n/d	n/d	n/d	1/1	
hsa-miR-590-3p	43752699-43752719	1	400	yes					1/5	
hsa-miR-4307	43752703-43752720	1	404	yes	n/d	n/d	n/d	n/d	1/1	
hsa-miR-186	43752709-43752731	1	410	yes	yes†		yes		3/5	
hsa-miR-548u	43752709-43752731	1	410	yes	n/d	n/d	n/d	n/d	1/1	
hsa-miR-3133	43752710-43752731	1	411	yes	n/d	n/d	n/d	n/d	1/1	

Appendix

MicroRNA	Genome coordinates	Reg.	Pos.	MR	TS	µT	DB	MC	#	Reference
hsa-miR-548l	43752710-43752731	1	411	yes		n/d		n/d	1/3	
hsa-miR-548n	43752711-43752732	1	412	yes		n/d		n/d	1/3	
hsa-miR-3121	43752712-43752733	1	413	yes	n/d	n/d	n/d	n/d	1/1	
hsa-miR-3123	43752714-43752731	1	415	yes	n/d	n/d	n/d	n/d	1/1	
hsa-miR-4311	43752717-43752733	1	418	yes	n/d	n/d	n/d	n/d	1/1	
hsa-miR-32*	43752720-43752740	1	421	yes	n/d	n/d	n/d		1/2	
hsa-miR-2053	43752722-43752744	1	423	yes		n/d	yes	n/d	2/3	
hsa-miR-569	43752723-43752743	1	424	yes	yes				2/5	
hsa-miR-452	43752725-43752746	1	426	yes					1/5	
hsa-miR-141	43752727-43752746	1	428	yes					1/5	
hsa-miR-1208	43752728-43752747	1	429	yes	yes	n/d	yes	n/d	3/3	
hsa-miR-106b*	43752728-43752749	1	429	yes	n/d	n/d			1/2	
hsa-miR-1278	43752728-43752749	1	429	yes		n/d		n/d	1/3	
hsa-miR-374b*	43752729-43752751	1	430	yes	n/d	n/d	n/d		1/2	
hsa-miR-323-3p	43752733-43752754	1	434	yes					1/5	
hsa-miR-543	43752734-43752755	1	435	yes	yes†				2/5	
hsa-miR-3143	43752736-43752756	1	437	yes	n/d	n/d	n/d	n/d	1/1	
hsa-miR-21*	43752743-43752764	1	444	yes	n/d	n/d	n/d		1/2	
hsa-miR-199b-5p	43752750-43752772	1	451	yes	yes†				2/5	
hsa-miR-508-5p	43752751-43752776	1	452	yes					1/5	
hsa-miR-199a-5p	43752753-43752772	1	454	yes	yes†				2/5	
hsa-miR-1825	43752755-43752772	1	456	yes	yes†	n/d		n/d	2/3	
hsa-miR-4276	43752756-43752772	1	457	yes	n/d	n/d	n/d	n/d	1/1	
hsa-miR-545	43752761-43752785	1	462	yes					1/5	
hsa-miR-4274	43752770-43752787	1	471	yes	n/d	n/d	n/d	n/d	1/1	
hsa-miR-150	43752783-43752805	1	484	yes	yes				2/5	
hsa-miR-543	43752807-43752813*	1	508		yes				1/5	
hsa-miR-1292	43752811-43752817*	1	512		yes	n/d		n/d	1/3	
hsa-miR-1205	43752813-43752833	1/NC	514	yes		n/d		n/d	1/3	
hsa-miR-339-5p	43752817-43752835	1/NC	518	yes	yes		yes		3/5	
hsa-miR-1274a	43752820-43752836	NC	521	yes	yes	n/d		n/d	2/3	
hsa-miR-877*	43752821-43752841	NC	522	yes	n/d	n/d	n/d		1/2	
hsa-miR-483-3p	43752826-43752849	NC	527	yes					1/5	
hsa-miR-1274b	43752829-43752835*	NC	530		yes	n/d		n/d	1/3	
hsa-miR-4279	43752832-43752847	NC	533	yes	n/d	n/d	n/d	n/d	1/1	
hsa-miR-1236	43752833-43752839+	NC	534		yes	n/d	yes	n/d	2/3	
hsa-miR-641	43752847-43752869	NC	548	yes					1/5	
hsa-miR-3148	43752857-43752879	NC	558	yes	n/d	n/d	n/d	n/d	1/1	
hsa-miR-4298	43752860-43752881	NC	561	yes	n/d	n/d	n/d	n/d	1/1	
hsa-miR-1302	43752861-43752881	NC	562	yes	yes	n/d		n/d	2/3	
hsa-miR-504	43752892-43752898*	NC	593		yes				1/5	
hsa-miR-1207-5p	43752904-43752910*	NC	605		yes	n/d		n/d	1/3	
hsa-miR-1285	43752907-43752913*	NC	608		yes	n/d		n/d	1/3	
hsa-miR-612	43752907-43752913*	NC	608		yes				1/5	
hsa-miR-874	43752927-43752933+	NC	628		yes		yes		2/5	
hsa-miR-146b-3p	43752927-43752933*	NC	628		yes				1/5	
hsa-miR-548a-3p	43752929-43752957$	NC	630		yes	yes			2/5	
hsa-miR-1279	43752942-43752948*	NC	643		yes	n/d		n/d	1/3	
hsa-miR-548e	43752951-43752957*	NC	652		yes	n/d		n/d	1/3	
hsa-miR-548f	43752951-43752957*	NC	652		yes	n/d		n/d	2/3	
hsa-miR-1323	43752952-43752958*	NC	653		yes	n/d		n/d	1/3	
hsa-miR-548o	43752952-43752958*	NC	653		yes	n/d		n/d	1/3	
hsa-miR-762	43752957-43752981	NC	658	yes	n/d	n/d	n/d	n/d	1/1	
hsa-miR-629	43752970-43752976*	NC	671		yes				1/5	
hsa-miR-24	43752986-43752992*	NC	687		yes				1/5	
hsa-miR-631	43752988-43753008	NC	689	yes	yes				2/5	
hsa-miR-511	43753005-43753025	NC	706	yes	yes				2/5	
hsa-miR-619	43753009-43753035	NC	710	yes					1/5	
hsa-miR-486-5p	43753016-43753040	NC	717	yes	yes				2/5	
hsa-miR-205	43753023-43753048‡	NC	724						0/5	Ye et al., 2008
hsa-miR-1274a	43753025-43753042	NC	726	yes	yes	n/d		n/d	2/3	
hsa-miR-1274b	43753035-43753041*	NC	736		yes	n/d		n/d	1/3	
hsa-miR-339-5p	43753035-43753041*	NC	736		yes				1/5	
hsa-miR-15b	43753069-43753094%	NC	770						0/5	Hua et al., 2006
hsa-miR-107	43753070-43753093%	NC	771						0/5	Hua et al., 2006
hsa-miR-3120	43753073-43753093	NC	774	yes	n/d	n/d	n/d	n/d	1/1	
hsa-miR-520g	43753073-43753099‡	NC	774						0/5	Ye et al., 2008
hsa-miR-127-5p	43753074-43753080*	NC	775		yes				1/5	
hsa-miR-17	43753075-43753098‡	NC	776						0/5	Ye et al., 2008
hsa-miR-520h	43753075-43753099‡	NC	776						0/5	Hua et al., 2006; Ye et al., 2008
hsa-miR-20b	43753076-43753098‡	NC	777						0/5	Hua et al., 2006; Ye et al., 2008
hsa-miR-186*	43753077-43753101	NC	778	yes	n/d	n/d	n/d		1/2	
hsa-miR-15a	43753077-43753094‡	NC	778						0/5	Ye et al., 2008
hsa-miR-330-3p	43753092-43753098*	NC	793		yes				1/5	Ye et al., 2008
hsa-miR-16	43753092-43753121‡	NC	793						0/5	Hua et al., 2006; Ye et al., 2008
hsa-miR-147	43753093-43753116‡	NC	794						0/5	Ye et al., 2008
hsa-miR-103	43753099-43753120	NC	800	yes	yes	yes			3/5	
hsa-miR-107	43753099-43753120	NC	800	yes	yes	yes			3/5	
hsa-miR-372	43753120-43753147‡	NC	821						0/5	Ye et al., 2008
hsa-miR-373	43753124-43753147‡	NC	825						0/5	Ye et al., 2008

Appendix

MicroRNA	Genome coordinates	Reg.	Pos.	MR	TS	µT	DB	MC	#	Reference
hsa-miR-637	43753132-43753138*	NC	833		yes				1/5	
hsa-miR-331-3p	43753135-43753141*	NC	836		yes				1/5	
hsa-miR-141*	43753137-43753157	NC	838	yes	n/d	n/d	n/d		1/2	
hsa-miR-378	43753141-43753163‡	NC	842						0/5	Ye et al., 2008
hsa-miR-34a	43753142-43753148*	NC	843		yes				1/5	
hsa-miR-34c-5p	43753142-43753148*	NC	843		yes				1/5	
hsa-miR-449a	43753142-43753148*	NC	843		yes				1/5	
hsa-miR-449b	43753142-43753148*	NC	843		yes				1/5	
hsa-miR-484	43753161-43753167*	NC	862		yes				1/5	
hsa-miR-4270	43753171-43753193	NC	872	yes	n/d	n/d	n/d	n/d	1/1	
hsa-miR-125a-3p	43753172-43753193	NC	873	yes	yes				2/5	
hsa-miR-612	43753187-43753209	NC	888	yes	yes				2/5	
hsa-miR-1285	43753189-43753209	NC	890	yes	yes	n/d		n/d	2/3	
hsa-miR-140-5p	43753189-43753217$	NC	890			yes			1/5	Ye et al., 2008
hsa-miR-942	43753220-43753242	NC	921	yes	yes				2/5	
hsa-miR-1236	43753233-43753239*	NC	934		yes	n/d		n/d	1/3	
hsa-miR-1236	43753238-43753244*	NC	939		yes	n/d		n/d	1/3	
hsa-miR-103-2*	43753240-43753263	NC	941	yes	n/d	n/d	n/d	n/d	1/1	
hsa-miR-593	43753240-43753246*	NC	941		yes				1/5	
hsa-miR-920	43753251-43753279$	NC	952		yes	yes			2/5	
hsa-miR-920	43753251-43753279$	NC	952			yes			1/5	
hsa-miR-920	43753251-43753279$	NC	952			yes			1/5	
hsa-miR-939	43753256-43753281	NC	957	yes	yes	yes			3/5	
hsa-miR-1308	43753265-43753271*	NC	966		yes	n/d		n/d	1/3	
hsa-miR-939	43753292-43753317	NC	993	yes	yes	yes			3/5	
hsa-miR-1255a	43753296-43753302*	NC	997		yes	n/d		n/d	1/3	
hsa-miR-1255b	43753296-43753302*	NC	997		yes	n/d		n/d	1/3	
hsa-miR-3179	43753299-43753320	NC	1000	yes	n/d	n/d	n/d	n/d	1/1	
hsa-miR-1224-5p	43753335-43753353	NC	1036	yes	n/d	yes	n/d		3/3	
hsa-miR-29b-2*	43753347-43753367	NC	1048	yes	n/d	n/d	n/d		1/2	
hsa-miR-1294	43753348-43753354*	NC	1049		yes			n/d	1/3	
hsa-miR-140-5p	43753352-43753369	NC	1053	yes	yes†	yes			3/5	
hsa-miR-637	43753362-43753385	NC	1063	yes	yes				2/5	
hsa-miR-4271	43753362-43753381	NC	1063	yes	n/d	n/d	n/d	n/d	1/1	
hsa-miR-342-3p	43753387-43753409	NC	1088	yes	yes				2/5	
hsa-miR-505	43753395-43753417	NC	1096	yes					1/5	
hsa-miR-493	43753399-43753421	NC	1100	yes	yes				2/5	
hsa-miR-377	43753401-43753407*	NC	1102		yes†				1/5	
hsa-miR-765	43753419-43753425*	NC	1120		yes				1/5	
hsa-miR-339-5p	43753449-43753471	NC	1150	yes	yes		yes		3/5	
hsa-miR-146b-3p	43753452-43753472	NC	1153	yes	yes				2/5	
hsa-miR-550	43753453-43753459*	NC	1154		yes				1/5	
hsa-miR-593	43753461-43753467*	NC	1162		yes				1/5	
hsa-miR-449a	43753464-43753486	NC	1165	yes					1/5	
hsa-miR-449b	43753464-43753486	NC	1165	yes					1/5	
hsa-miR-874	43753466-43753472+	NC	1167		yes		yes		2/5	
hsa-miR-768-3p	43753486-43753492*	NC/2	1187	n/d	yes		n/d		1/3	
hsa-miR-141	43753496-43753517	2	1197	yes					1/5	
hsa-miR-200a	43753496-43753517	2	1197	yes					1/5	
hsa-miR-3163	43753500-43753521	2	1201	yes	n/d	n/d	n/d	n/d	1/1	
hsa-miR-371-3p	43753509-43753531	2	1210	yes					1/5	
hsa-miR-624*	43753510-43753531	2	1211	yes	n/d	n/d	n/d		1/2	
hsa-miR-126	43753511-43753531	2	1212	yes				yes	2/5	Liu et al., 2009
hsa-miR-1911	43753511-43753532	2	1212	yes	n/d	n/d		n/d	1/2	
hsa-miR-548k	43753513-43753533	2	1214	yes			n/d	n/d	1/3	
hsa-miR-3145	43753517-43753542	2	1218	yes	n/d	n/d	n/d	n/d	1/1	
hsa-miR-410	43753517-43753523*	2	1218		yes†				1/5	
hsa-miR-656	43753521-43753541	2	1222	yes					1/5	
hsa-miR-3154	43753527-43753548	2	1228	yes	n/d	n/d	n/d	n/d	1/1	
hsa-miR-3201	43753527-43753543	2	1228	yes	n/d	n/d	n/d	n/d	1/1	
hsa-miR-548t	43753528-43753548	2	1229	yes	n/d	n/d	n/d	n/d	1/1	
hsa-miR-583	43753529-43753549	2	1230	yes					1/5	
hsa-miR-576-5p	43753536-43753557	2	1237	yes	yes†		yes		3/5	
hsa-miR-513a-3p	43753539-43753561	2	1240	yes		n/d		n/d	1/3	
hsa-miR-4263	43753540-43753557	2	1241	yes	n/d	n/d	n/d	n/d	1/1	
hsa-miR-29b-1*	43753545-43753567	2	1246	yes	n/d	n/d	n/d		1/2	
hsa-miR-452	43753547-43753568	2	1248	yes	yes†		yes		3/5	
hsa-miR-4307	43753547-43753567	2	1248	yes	n/d	n/d	n/d	n/d	1/1	
hsa-miR-2052	43753548-43753565	2	1249	yes	n/d	n/d		n/d	1/2	
hsa-miR-451	43753548-43753569	2	1249	yes					1/5	
hsa-miR-582-3p	43753548-43753569	2	1249	yes					1/5	
hsa-miR-1208	43753550-43753569	2	1251	yes		n/d		n/d	1/3	
hsa-miR-548g	43753550-43753570	2	1251	yes		n/d		n/d	1/3	
hsa-miR-943	43753550-43753570	2	1251	yes					1/5	
hsa-miR-183*	43753554-43753575	2	1255	yes	n/d	n/d	n/d		1/2	
hsa-miR-302b*	43753561-43753582	2	1262	yes	n/d	n/d	n/d		1/2	
hsa-miR-302d*	43753561-43753582	2	1262	yes	n/d	n/d	n/d		1/2	
hsa-miR-130b*	43753564-43753584	2	1265	yes	n/d	n/d	n/d		1/2	
hsa-miR-24-1*	43753568-43753589	2	1269	yes	n/d	n/d	n/d		1/2	
hsa-miR-24-2*	43753568-43753589	2	1269	yes	n/d	n/d	n/d		1/2	
hsa-miR-548d-3p	43753575-43753594	2	1276	yes	yes†		yes		3/5	
hsa-miR-548x	43753576-43753594	2	1277	yes	n/d	n/d	n/d	n/d	1/1	

Appendix

MicroRNA	Genome coordinates	Reg.	Pos.	MR	TS	µT	DB	MC	#	Reference
hsa-miR-875-3p	43753580-43753600	2	1281	yes					1/5	
hsa-miR-200c	43753582-43753604	2	1283	yes	yes†				2/5	
hsa-miR-200b	43753583-43753604	2	1284	yes	yes†				2/5	McArthur et al., 2011
hsa-miR-1278	43753583-43753604	2	1284	yes		n/d		n/d	1/3	
hsa-miR-4251	43753584-43753600	2	1285	yes	n/d	n/d	n/d		1/1	
hsa-miR-429	43753585-43753604	2	1286	yes	yes†				2/5	
hsa-miR-186	43753587-43753608	2	1288	yes					1/5	
hsa-miR-16-2*	43753595-43753618	2	1296	yes	n/d	n/d	n/d		1/2	
hsa-miR-656	43753596-43753616	2	1297	yes		yes			2/5	
hsa-miR-195*	43753596-43753618	2	1297	yes	n/d	n/d			1/2	
hsa-miR-559	43753603-43753623	2	1304	yes					1/5	
hsa-miR-203	43753604-43753625	2	1305	yes	yes†				2/5	
hsa-miR-653	43753606-43753628	2	1307	yes					1/5	
hsa-miR-219-1-3p	43753608-43753629	2	1309	yes					1/5	
hsa-miR-3121	43753611-43753632	2	1312	yes	n/d	n/d	n/d	n/d	1/1	
hsa-miR-633	43753611-43753632	2	1312	yes					1/5	
hsa-miR-216a	43753619-43753640	2	1320	yes					1/5	
hsa-miR-3133	43753624-43753647	2	1325	yes	n/d	n/d	n/d	n/d	1/1	
hsa-miR-548p	43753630-43753651	2	1331	yes	yes	n/d	yes	n/d	3/3	
hsa-miR-3121	43753650-43753670	2	1351	yes	n/d	n/d	n/d	n/d	1/1	
hsa-miR-150*	43753657-43753676	2	1358	yes	n/d	n/d	n/d		1/2	
hsa-miR-548d-3p	43753658-43753682	2	1359	yes	yes		yes		3/5	
hsa-miR-1537	43753659-43753680	2	1360	yes	n/d	n/d		n/d	1/2	
hsa-miR-300	43753665-43753686	2	1366	yes	yes				2/5	
hsa-miR-381	43753665-43753686	2	1366	yes	yes				2/5	
hsa-miR-1283	43753665-43753686	2	1366	yes		n/d		n/d	1/3	
hsa-let-7f-1*	43753666-43753687	2	1367	yes	n/d	n/d	n/d		1/2	
hsa-let-7b*	43753667-43753687	2	1368	yes	n/d	n/d	n/d		1/2	
hsa-miR-1284	43753667-43753688	2	1368	yes		n/d		n/d	1/3	
hsa-let-7a*	43753668-43753687	2	1369	yes	n/d	n/d	n/d		1/2	
hsa-miR-590-3p	43753674-43753694	2	1375	yes	yes				2/5	
hsa-miR-494	43753681-43753702	2	1382	yes	yes				2/5	
hsa-miR-4261	43753688-43753703	2	1389	yes	n/d	n/d	n/d	n/d	1/1	
hsa-miR-571	43753688-43753709	2	1389	yes					1/5	
hsa-miR-765	43753691-43753716	2	1392	yes					1/5	
hsa-miR-185	43753695-43753716	2	1396	yes	yes		yes		3/5	
hsa-miR-4270	43753700-43753719	2	1401	yes	n/d	n/d	n/d	n/d	1/1	
hsa-miR-4306	43753700-43753716	2	1401	yes	n/d	n/d	n/d	n/d	1/1	
hsa-miR-3118	43753710-43753734	2	1411	yes	n/d	n/d	n/d	n/d	1/1	
hsa-miR-134	43753711-43753734	2	1412	yes	yes		yes		3/5	
hsa-miR-943	43753712-43753733	2	1413	yes	yes			yes	3/5	
hsa-miR-1278	43753712-43753735	2	1413	yes		n/d		n/d	1/3	
hsa-miR-512-3p	43753713-43753735	2	1414	yes					1/5	
hsa-miR-26a	43753724-43753747	2	1425	yes					1/5	
hsa-miR-26b*	43753730-43753751	2	1431	yes	n/d	n/d	n/d		1/2	
hsa-miR-124*	43753731-43753751	2	1432	yes	n/d	n/d	n/d		1/2	
hsa-miR-429	43753734-43753756	2	1435	yes					1/5	
hsa-miR-889	43753736-43753757	2	1437	yes					1/5	
hsa-miR-548p	43753747-43753766	2	1448	yes	yes†	n/d	yes	n/d	3/3	
hsa-miR-3152	43753751-43753771	2	1452	yes	n/d	n/d	n/d	n/d	1/1	
hsa-miR-138	43753752-43753776	2	1453	yes					1/5	
hsa-miR-3160	43753756-43753778	2	1457	yes	n/d	n/d	n/d	n/d	1/1	
hsa-miR-650	43753765-43753786	2	1466	yes					1/5	
hsa-miR-574-5p	43753767-43753773*	2	1468		yes				1/5	
hsa-miR-149*	43753769-43753787	2	1470	yes	n/d	n/d	n/d		1/2	
hsa-miR-638	43753771-43753794	2	1472	yes	yes				2/5	
hsa-miR-939	43753780-43753804	2	1481	yes	yes	yes			3/5	
hsa-miR-505*	43753783-43753802	2	1484	yes	n/d	n/d	n/d		1/2	
hsa-miR-542-5p	43753783-43753807*	2	1484		yes				1/5	
hsa-miR-3150	43753786-43753805	2	1487	yes	n/d	n/d	n/d	n/d	1/1	
hsa-miR-1231	43753790-43753810	2	1491	yes		n/d		n/d	1/3	
hsa-miR-409-3p	43753793-43753814	2	1494	yes					1/5	
hsa-miR-219-2-3p	43753794-43753815	2	1495	yes					1/5	
hsa-miR-206	43753795-43753816	2	1496	yes	yes†		yes		3/5	
hsa-miR-1	43753795-43753816	2	1496	yes	yes†				2/5	
hsa-miR-613	43753796-43753816	2	1497	yes	yes†	yes			3/5	
hsa-miR-607	43753802-43753822	2	1503	yes	yes				2/5	
hsa-miR-3171	43753817-43753842	2	1518	yes	n/d	n/d	n/d	n/d	1/1	
hsa-miR-568	43753821-43753840	2	1522	yes					1/5	
hsa-miR-1279	43753822-43753838	2	1523	yes		n/d		n/d	1/3	
hsa-miR-576-3p	43753826-43753843	2	1527	yes	yes				2/5	
hsa-miR-1277	43753826-43753847	2	1527	yes		n/d		n/d	1/3	
hsa-miR-567	43753827-43753849	2	1528	yes					1/5	
hsa-miR-297	43753828-43753850	2	1529	yes					1/5	
hsa-miR-3149	43753829-43753851	2	1530	yes	n/d	n/d	n/d	n/d	1/1	
hsa-miR-1265	43753829-43753850	2	1530	yes		n/d		n/d	1/3	
hsa-miR-633	43753832-43753855	2	1533	yes					1/5	
hsa-miR-144	43753834-43753853	2	1535	yes					1/5	
hsa-miR-656	43753839-43753860	2	1540	yes					1/5	
hsa-miR-548s	43753845-43753867	2	1546	yes	n/d	n/d	n/d	n/d	1/1	
hsa-miR-192*	43753845-43753866	2	1546	yes	n/d	n/d	n/d		1/2	
hsa-miR-744*	43753848-43753869	2	1549	yes	n/d	n/d	n/d		1/2	

Appendix

MicroRNA	Genome coordinates	Reg.	Pos.	MR	TS	µT	DB	MC	#	Reference
hsa-miR-382	43753849-43753870	2	1550	yes					1/5	
hsa-miR-300	43753852-43753874	2	1553	yes	yes				2/5	
hsa-miR-381	43753852-43753874	2	1553	yes	yes				2/5	
hsa-miR-655	43753856-43753876	2	1557	yes					1/5	
hsa-miR-7-1*	43753859-43753879	2	1560	yes	n/d	n/d	n/d		1/2	
hsa-miR-7-2*	43753859-43753879	2	1560	yes	n/d	n/d	n/d		1/2	
hsa-miR-329	43753861-43753882	2	1562	yes	yes				2/5	
hsa-miR-362-3p	43753861-43753882	2	1562	yes	yes				2/5	
hsa-miR-603	43753861-43753882	2	1562	yes	yes				2/5	
hsa-miR-466	43753861-43753883	2	1562	yes	n/d	n/d	n/d	n/d	1/1	
hsa-miR-377	43753861-43753882	2	1562	yes					1/5	
hsa-let-7b*	43753864-43753885	2	1565	yes	n/d	n/d	n/d		1/2	
hsa-let-7f-1*	43753864-43753885	2	1565	yes	n/d	n/d	n/d		1/2	
hsa-let-7f-2*	43753864-43753885	2	1565	yes	n/d	n/d	n/d		1/2	
hsa-let-7a*	43753865-43753885	2	1566	yes	n/d	n/d	n/d		1/2	
hsa-miR-1284	43753865-43753886	2	1566	yes		n/d		n/d	1/3	
hsa-miR-410	43753865-43753886	2	1566	yes					1/5	
hsa-miR-374a	43753867-43753889	2	1568	yes					1/5	
hsa-miR-656	43753869-43753890	2	1570	yes					1/5	
hsa-miR-30c	43753882-43753904	2	1583	yes					1/5	
hsa-miR-494	43753882-43753903	2	1583	yes					1/5	
hsa-miR-302a*	43753883-43753905	2	1584	yes	n/d	n/d	n/d		1/2	
hsa-miR-30a	43753883-43753904	2	1584	yes					1/5	
hsa-miR-30b	43753883-43753904	2	1584	yes					1/5	
hsa-miR-30d	43753883-43753904	2	1584	yes					1/5	
hsa-miR-30e	43753883-43753904	2	1584	yes					1/5	
hsa-let-7b*	43753888-43753909	2	1589	yes	n/d	n/d	n/d		1/2	
hsa-let-7f-1*	43753888-43753909	2	1589	yes	n/d	n/d	n/d		1/2	
hsa-let-7f-2*	43753888-43753909	2	1589	yes	n/d	n/d	n/d		1/2	
hsa-miR-1284	43753888-43753910	2	1589	yes		n/d		n/d	1/3	
hsa-let-7a*	43753889-43753909	2	1590	yes	n/d	n/d	n/d		1/2	
hsa-miR-130a*	43753895-43753916	2	1596	yes	n/d	n/d	n/d		1/2	
hsa-miR-23a	43753896-43753916	2	1597	yes					1/5	
hsa-miR-23b	43753896-43753916	2	1597	yes					1/5	
hsa-miR-15b*	43753897-43753919	2	1598	yes	n/d	n/d	n/d		1/2	
hsa-miR-34b	43753897-43753918	2	1598	yes					1/5	
hsa-miR-2115*	43753901-43753923	2	1602	yes	n/d	n/d	n/d	n/d	1/1	
hsa-miR-3074	43753902-43753925	2	1603	yes	n/d	n/d	n/d	n/d	1/1	
hsa-miR-374a*	43753902-43753925	2	1603	yes	n/d	n/d	n/d		1/2	
hsa-miR-383	43753902-43753924	2	1603	yes					1/5	
hsa-miR-361-5p	43753903-43753924	2	1604	yes	yes†	yes	yes	yes	5/5	
hsa-miR-138-2*	43753909-43753930	2	1610	yes	n/d	n/d	n/d		1/2	
hsa-miR-654-3p	43753909-43753934	2	1610	yes					1/5	
hsa-miR-934	43753909-43753932	2	1610	yes					1/5	
hsa-miR-590-3p	43753910-43753929	2	1611	yes					1/5	
hsa-miR-106a*	43753917-43753938	2	1618	yes	n/d	n/d	n/d		1/2	
hsa-miR-7-2*	43753922-43753946	2	1623	yes	n/d	n/d	n/d		1/2	
hsa-miR-3133	43753929-43753950	2	1630	yes	n/d	n/d	n/d	n/d	1/1	
hsa-miR-548x	43753929-43753949	2	1630	yes	n/d	n/d	n/d	n/d	1/1	
hsa-miR-548d-3p	43753929-43753950	2	1630	yes					1/5	
hsa-miR-570	43753929-43753949	2	1630	yes					1/5	
hsa-miR-3163	43753930-43753952	2	1631	yes	n/d	n/d	n/d	n/d	1/1	
hsa-miR-410	43753931-43753954	2	1632	yes					1/5	
hsa-miR-577	43753933-43753954	2	1634	yes					1/5	
hsa-miR-340	43753934-43753953	2	1635	yes					1/5	
hsa-miR-376c	43753936-43753956	2	1637	yes					1/5	
hsa-miR-448	43753936-43753957	2	1637	yes					1/5	
hsa-miR-466	43753937-43753957	2	1638	yes	n/d	n/d	n/d	n/d	1/1	
hsa-miR-2052	43753942-43753962	2	1643	yes	n/d	n/d		n/d	1/2	
hsa-miR-129-5p	43753942-43753963	2	1643	yes					1/5	
hsa-miR-2116	43753943-43753962	2	1644	yes	n/d	n/d	n/d	n/d	1/1	
hsa-miR-4307	43753946-43753967	2	1647	yes	n/d	n/d	n/d	n/d	1/1	
hsa-miR-19a*	43753946-43753967	2	1647	yes	n/d	n/d	n/d		1/2	
hsa-miR-19b-1*	43753947-43753967	2	1648	yes	n/d	n/d	n/d		1/2	
hsa-miR-19b-2*	43753948-43753967	2	1649	yes	n/d	n/d	n/d		1/2	
hsa-miR-578	43753952-43753972	2	1653	yes	yes†		yes		3/5	
hsa-miR-138-2*	43753958-43753979	2	1659	yes	n/d	n/d	n/d		1/2	
hsa-miR-136	43753959-43753981	2	1660	yes					1/5	
hsa-miR-942	43753963-43753984	2	1664	yes					1/5	
hsa-miR-2117	43753964-43753984	2	1665	yes	n/d	n/d	n/d	n/d	1/1	
hsa-miR-1277	43753970-43753991	2	1671	yes		n/d		n/d	1/3	
hsa-miR-567	43753971-43753993	2	1672	yes					1/5	
hsa-miR-3149	43753973-43753995	2	1674	yes	n/d	n/d	n/d	n/d	1/1	
hsa-miR-297	43753975-43753994	2	1676	yes					1/5	
hsa-miR-32*	43753978-43753999	2	1679	yes	n/d	n/d	n/d		1/2	
hsa-miR-185	43753985-43754006	2	1686	yes	yes†		yes		3/5	
hsa-miR-3173	43753986-43754008	2	1687	yes	n/d	n/d	n/d	n/d	1/1	
hsa-miR-4306	43753990-43754006	2	1691	yes	n/d	n/d	n/d	n/d	1/1	
hsa-miR-583	43753991-43754011	2	1692	yes					1/5	
hsa-miR-3143	43753995-43754019	2	1696	yes	n/d	n/d	n/d	n/d	1/1	
hsa-miR-551b*	43753995-43754017	2	1696	yes	n/d	n/d	n/d		1/2	
hsa-miR-548n	43753997-43754018	2	1698	yes		n/d		n/d	1/3	

Appendix

MicroRNA	Genome coordinates	Reg.	Pos.	MR	TS	µT	DB	MC	#	Reference
hsa-miR-3163	43753998-43754020	2	1699	yes	n/d	n/d	n/d	n/d	1/1	
hsa-miR-4282	43754001-43754018	2	1702	yes	n/d	n/d	n/d		1/1	
hsa-miR-656	43754003-43754023	2	1704	yes		yes			2/5	
hsa-miR-16-2*	43754004-43754025	2	1705	yes	n/d	n/d	n/d		1/2	
hsa-miR-195*	43754004-43754025	2	1705	yes	n/d	n/d	n/d		1/2	
hsa-miR-3065-5p	43754006-43754029	2	1707	yes	n/d	n/d	n/d	n/d	1/1	
hsa-miR-495	43754007-43754028	2	1708	yes	yes				2/5	
hsa-miR-7-1*	43754007-43754028	2	1708	yes	n/d	n/d	n/d		1/2	
hsa-miR-7-2*	43754007-43754028	2	1708	yes	n/d	n/d	n/d		1/2	
hsa-miR-3161	43754010-43754032	2	1711	yes	n/d	n/d	n/d	n/d	1/1	
hsa-miR-570	43754010-43754031	2	1711	yes					1/5	
hsa-miR-224*	43754011-43754035	2	1712	yes	n/d	n/d	n/d	n/d	1/1	
hsa-miR-522	43754011-43754035	2	1712	yes					1/5	
hsa-miR-577	43754019-43754039	2	1720	yes					1/5	
hsa-miR-323b-3p	43754023-43754044	2	1724	yes	n/d	n/d	n/d	n/d	1/1	
hsa-miR-593*	43754024-43754048	2	1725	yes	n/d	n/d	n/d		1/2	
hsa-miR-767-5p	43754025-43754047	2	1726	yes					1/5	
hsa-miR-3065-3p	43754027-43754049	2	1728	yes	n/d	n/d	n/d	n/d	1/1	
hsa-miR-424	43754027-43754048	2	1728	yes					1/5	
hsa-miR-29b	43754028-43754048	2	1729	yes	yes†	yes	yes	yes	5/5	
hsa-miR-21*	43754028-43754048	2	1729	yes	n/d	n/d	n/d		1/2	
hsa-miR-222*	43754028-43754053	2	1729	yes	n/d	n/d	n/d		1/2	
hsa-miR-29a	43754029-43754048	2	1730	yes	yes†	yes	yes		4/5	
hsa-miR-29c	43754029-43754048	2	1730	yes	yes†	yes	yes		4/5	
hsa-miR-3129	43754029-43754052	2	1730	yes	n/d	n/d	n/d	n/d	1/1	
hsa-miR-148a	43754030-43754052	2	1731	yes					1/5	
hsa-miR-148b	43754030-43754052	2	1731	yes					1/5	
hsa-miR-936	43754031-43754052	2	1732	yes					1/5	
hsa-miR-101	43754032-43754053	2	1733	yes					1/5	
hsa-miR-199b-3p	43754033-43754052	2	1734	yes				n/d	1/4	
hsa-miR-199a-3p	43754033-43754052	2	1734	yes					1/5	
hsa-miR-144	43754034-43754053	2	1735	yes					1/5	
hsa-miR-562	43754036-43754052	2	1737	yes					1/5	
hsa-miR-651	43754041-43754063	2	1742	yes					1/5	
hsa-miR-1261	43754043-43754060	2	1744	yes		n/d		n/d	1/3	
hsa-miR-556-3p	43754044-43754065	2	1745	yes	yes†				2/5	
hsa-miR-944	43754047-43754068	2	1748	yes	yes†		yes		3/5	
hsa-miR-1975	43754047-43754074	2	1748	yes		n/d	yes	n/d	2/3	
hsa-miR-126*	43754047-43754067	2	1748	yes	n/d	n/d	n/d		1/2	
hsa-miR-590-3p	43754047-43754068	2	1748	yes					1/5	
hsa-miR-2355	43754053-43754075	2	1754	yes	n/d	n/d	n/d	n/d	1/1	
hsa-miR-188-3p	43754054-43754074	2	1755	yes					1/5	
hsa-miR-4286	43754058-43754074	2	1759	yes	n/d	n/d	n/d		1/2	
hsa-miR-889	43754065-43754085	2	1766	yes	yes†		yes		3/5	
hsa-miR-2053	43754066-43754088	2	1767	yes		n/d	yes	n/d	2/3	
hsa-miR-569	43754067-43754087	2	1768	yes	yes†				2/5	
hsa-miR-153	43754086-43754110	2	1787	yes					1/5	
hsa-miR-548a-3p	43754095-43754115	2	1796	yes	yes†	yes			3/5	
hsa-miR-548e	43754095-43754115	2	1796	yes	yes†	n/d		n/d	2/3	
hsa-miR-548x	43754096-43754116	2	1797	yes	n/d	n/d	n/d		1/1	
hsa-miR-548f	43754097-43754115	2	1798	yes	yes†			n/d	2/3	
hsa-miR-516a-5p	43754101-43754121	2	1802	yes					1/5	
hsa-miR-191	43754105-43754130	2	1806	yes					1/5	
hsa-miR-190	43754118-43754139	2	1819	yes					1/5	
hsa-miR-190b	43754119-43754139	2	1820	yes					1/5	
hsa-miR-548c-3p	43754125-43754146	2	1826	yes	yes†				2/5	
hsa-miR-548n	43754125-43754146	2	1826	yes		n/d		n/d	1/3	
hsa-miR-29a*	43754144-43754168	2	1845	yes	n/d	n/d	n/d		1/2	
hsa-miR-138-2*	43754145-43754166	2	1846	yes	n/d	n/d	n/d		1/2	
hsa-miR-876-5p	43754145-43754166	2	1846	yes					1/5	
hsa-miR-223*	43754147-43754168	2	1848	yes	n/d	n/d	n/d		1/2	
hsa-miR-654-3p	43754151-43754172	2	1852	yes					1/5	
hsa-miR-651	43754158-43754179	2	1859	yes					1/5	
hsa-miR-3140	43754170-43754191	2	1871	yes	n/d	n/d	n/d	n/d	1/1	
hsa-miR-586	43754170-43754193	2	1871	yes					1/5	
hsa-miR-3119	43754173-43754192	2	1874	yes	n/d	n/d	n/d	n/d	1/1	
hsa-miR-155	43754173-43754195	2	1874	yes					1/5	
hsa-miR-105	43754174-43754196	2	1875	yes					1/5	
hsa-miR-106a	43754174-43754196	2	1875	yes					1/5	
hsa-miR-17	43754174-43754196	2	1875	yes					1/5	
hsa-miR-20a	43754174-43754196	2	1875	yes					1/5	
hsa-miR-20b	43754174-43754196	2	1875	yes					1/5	
hsa-miR-93	43754174-43754196	2	1875	yes					1/5	
hsa-miR-4307	43754175-43754193	2	1876	yes	n/d	n/d	n/d	n/d	1/1	
hsa-miR-526b*	43754175-43754196	2	1876	yes	n/d	n/d	n/d		1/2	
hsa-miR-519d	43754175-43754196	2	1876	yes					1/5	
hsa-miR-106b	43754176-43754196	2	1877	yes					1/5	
hsa-miR-3163	43754177-43754198	2	1878	yes	n/d	n/d	n/d	n/d	1/1	
hsa-miR-300	43754178-43754200	2	1879	yes					1/5	
hsa-miR-524-5p	43754178-43754199	2	1879	yes					1/5	
hsa-let-7b*	43754180-43754201	2	1881	yes	n/d	n/d	n/d		1/2	
hsa-miR-520d-5p	43754180-43754199	2	1881	yes					1/5	

Appendix

MicroRNA	Genome coordinates	Reg.	Pos.	MR	TS	µT	DB	MC	#	Reference
hsa-miR-655	43754181-43754202	2	1882	yes	yes†		yes		3/5	
hsa-let-7a*	43754181-43754201	2	1882	yes	n/d	n/d	n/d		1/2	
hsa-miR-1283	43754182-43754200	2	1883	yes		n/d		n/d	1/3	
hsa-miR-381	43754182-43754200	2	1883	yes					1/5	
hsa-let-7f-1*	43754183-43754201	2	1884	yes	n/d	n/d	n/d		1/2	
hsa-miR-302d*	43754183-43754205	2	1884	yes	n/d	n/d	n/d		1/2	
hsa-miR-369-3p	43754183-43754203	2	1884	yes					1/5	
hsa-miR-302b*	43754185-43754205	2	1886	yes	n/d	n/d	n/d		1/2	
hsa-miR-10a*	43754189-43754210	2	1890	yes	n/d	n/d	n/d		1/2	
hsa-miR-148a*	43754189-43754210	2	1890	yes	n/d	n/d	n/d		1/2	
hsa-miR-16-1*	43754193-43754217	2	1894	yes	n/d	n/d	n/d		1/2	
hsa-miR-183*	43754195-43754216	2	1896	yes	n/d	n/d	n/d		1/2	
hsa-miR-2115*	43754198-43754219	2	1899	yes	n/d	n/d	n/d	n/d	1/1	
hsa-miR-383	43754199-43754220	2	1900	yes	yes†				2/5	
hsa-miR-361-5p	43754199-43754220	2	1900	yes					1/5	

7.5 Predicted microRNA 361-5p targets

Target predictions were from the following web services: microRNA.org (MR; Betel *et al.*, 2010), TargetScan (TS; Friedman *et al.*, 2009), DIANA-microT (µT; Maragkakis *et al.*, 2009), miRDB (DB; Wang, 2008), and MicroCosm (MC; Griffiths-Jones *et al.*, 2008). Results were pooled and converted to uniform gene identifiers (Entrez) using the DAVID web service (Huang *et al.*, 2008). For each gene the particular algorithms and the total number of algorithms predicting it (#) are indicated. Genes acting in the VEGF pathway according to KEGG (Kanehisa *et al.*, 2010) are indicated (KEGG). The list is restricted to target candidates that were predicted by at least two different algorithms.

Symbol	Gene name	Entrez	MR	TS	µT	DB	MC	#	KEGG
BCORL1	BCL6 co-repressor-like 1	63035	yes	yes	yes	yes	yes	5	
POLR3G	polymerase (RNA) III (DNA directed) polypeptide G	10622	yes	yes	yes	yes	yes	5	
RANBP17	RAN binding protein 17	64901	yes	yes	yes	yes	yes	5	
RRAGB	Ras-related GTP binding B	10325	yes	yes	yes	yes	yes	5	
VEGFA	vascular endothelial growth factor A	7422	yes	yes	yes	yes	yes	5	yes
ARGLU1	arginine and glutamate rich 1	55082	yes	yes		yes	yes	4	
ARMC8	armadillo repeat containing 8	25852	yes	yes	yes	yes		4	
ATPAF1	ATP synthase mitochondrial F1 complex assembly	64756	yes	yes		yes	yes	4	
C18ORF34	chromosome 18 open reading frame 34	374864	yes	yes		yes	yes	4	
C9ORF150	chromosome 9 open reading frame 150	286343	yes	yes		yes	yes	4	
CAMK2D	calcium/calmodulin-dependent protein kinase II delta	817	yes	yes	yes	yes		4	
CPOX	coproporphyrinogen oxidase	1371	yes	yes		yes	yes	4	
CSMD1	CUB and Sushi multiple domains 1	64478	yes	yes		yes	yes	4	
CSTF1	cleavage stimulation factor, 3' pre-RNA, subunit 1,	1477	yes	yes		yes	yes	4	
DCTN6	dynactin 6	10671	yes		yes	yes	yes	4	
ELOVL7	ELOVL family member 7, elongation of long chain	79993	yes	yes		yes	yes	4	
FAM122C	family with sequence similarity 122C	159091	yes	yes		yes	yes	4	
GLRB	glycine receptor, beta	2743	yes	yes	yes		yes	4	
GNB4	guanine nucleotide binding protein (G protein), beta	59345	yes		yes	yes	yes	4	
GOLGA4	golgi autoantigen, golgin subfamily a, 4	2803	yes		yes	yes	yes	4	
MAP2	microtubule-associated protein 2	4133	yes	yes		yes	yes	4	
MTRR	5-methyltetrahydrofolate-homocysteine	4552	yes	yes	yes	yes		4	
NFE2	nuclear factor (erythroid-derived) 2, 45kDa	4778	yes	yes			yes	4	
OGN	osteoglycin	4969	yes	yes		yes	yes	4	
PRICKLE2	prickle homolog 2 (Drosophila)	166336	yes	yes		yes	yes	4	
RBM16	RNA binding motif protein 16	22828	yes	yes		yes	yes	4	
SEC31A	SEC31 homolog A (S. cerevisiae)	22872	yes		yes	yes	yes	4	
SH3BGRL2	SH3 domain binding glutamic acid-rich protein like 2	83699	yes		yes	yes	yes	4	
TFAP2B	transcription factor AP-2 beta (activating enhancer	7021	yes	yes		yes	yes	4	
TSR1	TSR1, 20S rRNA accumulation, homolog (S.	55720	yes		yes	yes	yes	4	
WNT7A	wingless-type MMTV integration site family, member	7476	yes	yes		yes	yes	4	
ABCA10	ATP-binding cassette, sub-family A (ABC1), member	10349	yes			yes	yes	3	
ABHD3	abhydrolase domain containing 3	171586	yes			yes	yes	3	
ACADSB	acyl-Coenzyme A dehydrogenase, short/branched	36	yes			yes	yes	3	
ADAMTS5	ADAM metallopeptidase with thrombospondin type 1	11096	yes			yes	yes	3	
ADCY2	adenylate cyclase 2 (brain)	108	yes	yes		yes		3	
ANKRD49	ankyrin repeat domain 49	54851	yes			yes	yes	3	
ARCN1	archain 1	372	yes	yes		yes		3	
BCAS1	breast carcinoma amplified sequence 1	8537	yes			yes	yes	3	
BMPR2	bone morphogenetic protein receptor, type II	659	yes	yes	yes			3	
C17ORF39	chromosome 17 open reading frame 39	79018	yes	yes		yes		3	
C1ORF77	chromosome 1 open reading frame 77	26097	yes		yes		yes	3	
C2ORF78	chromosome 2 open reading frame 78	388960	yes			yes	yes	3	
C3ORF17	chromosome 3 open reading frame 17	25871	yes		yes	yes		3	
C5ORF51	chromosome 5 open reading frame 51	285636	yes		yes	yes		3	
C6ORF58	chromosome 6 open reading frame 58	352999	yes			yes	yes	3	
C7ORF23	chromosome 7 open reading frame 23	79161	yes		yes	yes		3	

Appendix

Symbol	Gene name	Entrez	MR	TS	µT	DB	MC	#	KEGG
CAPN6	calpain 6	827	yes	yes		yes		3	
CBLN1	cerebellin 1 precursor	869	yes			yes	yes	3	
CCDC58	coiled-coil domain containing 58	131076	yes			yes	yes	3	
CCNYL1	cyclin Y-like 1	151195	yes			yes	yes	3	
CCT4	chaperonin containing TCP1, subunit 4 (delta)	10575	yes			yes	yes	3	
CD177	CD177 molecule	57126	yes	yes		yes		3	
CKS1B	CDC28 protein kinase regulatory subunit 1B	1163	yes			yes	yes	3	
CLDN22	claudin 22	53842	yes	yes		yes		3	
CMTM4	CKLF-like MARVEL transmembrane domain	146223	yes			yes	yes	3	
COQ2	coenzyme Q2 homolog, prenyltransferase (yeast)	27235	yes			yes	yes	3	
COX7A2	cytochrome c oxidase subunit VIIa polypeptide 2	1347	yes			yes	yes	3	
CRY1	cryptochrome 1 (photolyase-like)	1407	yes			yes	yes	3	
CYHR1	cysteine/histidine-rich 1	50626	yes			yes	yes	3	
DDX3X	DEAD (Asp-Glu-Ala-Asp) box polypeptide 3, X-linked	1654	yes	yes		yes	yes	3	
DEXI	dexamethasone-induced transcript	28955	yes			yes		3	
DPP10	dipeptidyl-peptidase 10	57628	yes	yes	yes			3	
DYNC1LI2	dynein, cytoplasmic 1, light intermediate chain 2	1783	yes		yes	yes		3	
EHF	ets homologous factor	26298	yes			yes		3	
EIF3A	eukaryotic translation initiation factor 3, subunit A	8661	yes			yes	yes	3	
ELL3	elongation factor RNA polymerase II-like 3	80237	yes			yes	yes	3	
ELMOD1	ELMO/CED-12 domain containing 1	55531	yes			yes	yes	3	
ERG	v-ets erythroblastosis virus E26 oncogene homolog	2078	yes	yes		yes		3	
FYTTD1	forty-two-three domain containing 1	84248	yes			yes	yes	3	
G6PC2	glucose-6-phosphatase, catalytic, 2	57818	yes			yes	yes	3	
GOT1	glutamic-oxaloacetic transaminase 1, soluble	2805	yes			yes	yes	3	
GPR155	G protein-coupled receptor 155	151556	yes	yes		yes		3	
GPRIN3	GPRIN family member 3	285513	yes		yes	yes		3	
GTPBP10	GTP-binding protein 10 (putative)	85865	yes		yes	yes		3	
HACE1	HECT domain and ankyrin repeat containing, E3	57531	yes			yes	yes	3	
HBS1L	HBS1-like (S. cerevisiae)	10767	yes	yes		yes		3	
HMMR	hyaluronan-mediated motility receptor (RHAMM)	3161	yes			yes	yes	3	
INTS8	integrator complex subunit 8	55656	yes			yes	yes	3	
IQCH	IQ motif containing H	64799	yes			yes	yes	3	
KIAA1524	KIAA1524	57650	yes			yes	yes	3	
LAD1	ladinin 1	3898	yes	yes		yes		3	
MCPH1	microcephalin 1	79648	yes			yes	yes	3	
MECP2	methyl CpG binding protein 2 (Rett syndrome)	4204	yes		yes	yes		3	
MET	met proto-oncogene (hepatocyte growth factor	4233	yes		yes	yes		3	
MRPL19	mitochondrial ribosomal protein L19	9801	yes		yes	yes		3	
MTUS1	mitochondrial tumor suppressor 1	57509	yes		yes	yes		3	
MTX2	metaxin 2	10651	yes			yes	yes	3	
NAALAD2	N-acetylated alpha-linked acidic dipeptidase 2	10003	yes			yes	yes	3	
NEBL	nebulette	10529	yes	yes	yes			3	
NPVF	neuropeptide VF precursor	64111	yes			yes	yes	3	
PARP11	poly (ADP-ribose) polymerase family, member 11	57097	yes	yes		yes		3	
PCDH11X	protocadherin 11 X-linked	27328	yes			yes	yes	3	
PDE4B	phosphodiesterase 4B, cAMP-specific	5142	yes	yes	yes			3	
PF4	platelet factor 4	5196	yes			yes	yes	3	
PGBD1	piggyBac transposable element derived 1	84547	yes			yes	yes	3	
PIGA	phosphatidylinositol glycan anchor biosynthesis, class	5277	yes	yes		yes		3	
PRKCB	protein kinase C, beta	5579	yes			yes	yes	3	yes
PRPF38A	PRP38 pre-mRNA processing factor 38 (yeast)	84950	yes			yes	yes	3	
RAB28	RAB28, member RAS oncogene family	9364	yes			yes	yes	3	
RAB3GAP1	RAB3 GTPase activating protein subunit 1 (catalytic)	22930	yes	yes		yes		3	
RAC1	ras-related C3 botulinum toxin substrate 1 (rho family,	5879	yes		yes	yes		3	yes
RCHY1	ring finger and CHY zinc finger domain containing 1	25898	yes			yes	yes	3	
RDH11	retinol dehydrogenase 11 (all-trans/9-cis/11-cis)	51109	yes		yes	yes		3	
RPGR	retinitis pigmentosa GTPase regulator	6103	yes			yes	yes	3	
SERP1	stress-associated endoplasmic reticulum protein 1	27230	yes			yes	yes	3	
SLC10A1	solute carrier family 10 (sodium/bile acid	6554	yes			yes	yes	3	
SLC1A3	solute carrier family 1 (glial high affinity glutamate	6507	yes	yes		yes		3	
SLC22A10	solute carrier family 22, member 10	387775	yes			yes	yes	3	
SMG1	SMG1 homolog, phosphatidylinositol 3-kinase-related	23049	yes		yes	yes		3	
SNCAIP	synuclein, alpha interacting protein	9627	yes		yes	yes		3	
SP1	Sp1 transcription factor	6667	yes	yes		yes		3	
STEAP2	six transmembrane epithelial antigen of the prostate 2	261729	yes		yes	yes		3	
STX7	syntaxin 7	8417	yes			yes	yes	3	
TAF5L	TAF5-like RNA polymerase II, p300/CBP-associated	27097	yes		yes	yes		3	
TC2N	tandem C2 domains, nuclear	123036	yes			yes	yes	3	
TFPI	tissue factor pathway inhibitor (lipoprotein-associated	7035	yes			yes	yes	3	
TPD52L3	tumor protein D52-like 3	89882	yes			yes	yes	3	
TRPC5	transient receptor potential cation channel, subfamily	7224	yes		yes	yes		3	
TSC1	tuberous sclerosis 1	7248	yes	yes	yes			3	
TYW3	tRNA-yW synthesizing protein 3 homolog (S.	127253	yes	yes	yes			3	
UBE2K	ubiquitin-conjugating enzyme E2K (UBC1 homolog,	3093	yes			yes	yes	3	
VBP1	von Hippel-Lindau binding protein 1	7411	yes			yes	yes	3	
WNT3	wingless-type MMTV integration site family, member	7473	yes		yes		yes	3	
XRCC4	X-ray repair complementing defective repair in	7518	yes			yes	yes	3	
YTHDF1	YTH domain family, member 1	54915	yes			yes	yes	3	
ZBTB40	zinc finger and BTB domain containing 40	9923	yes			yes	yes	3	
ZDHHC13	zinc finger, DHHC-type containing 13	54503	yes			yes	yes	3	
ZFAND1	zinc finger, AN1-type domain 1	79752	yes	yes		yes		3	

Appendix

Symbol	Gene name	Entrez	MR	TS	µT	DB	MC	#	KEGG
ZNF148	zinc finger protein 148	7707	yes	yes		yes		3	
ZNF238	zinc finger protein 238	10472	yes	yes		yes		3	
ZNF804A	zinc finger protein 804A	91752	yes			yes	yes	3	
AAK1	AP2 associated kinase 1	22848		yes		yes		2	
ABHD2	abhydrolase domain containing 2	11057		yes			yes	2	
ABL1	c-abl oncogene 1, receptor tyrosine kinase	25	yes			yes		2	
ACTG1	actin, gamma 1	71	yes				yes	2	
ADC	arginine decarboxylase	113451	yes				yes	2	
ADD3	adducin 3 (gamma)	120	yes			yes		2	
ADHFE1	alcohol dehydrogenase, iron containing, 1	137872	yes				yes	2	
AEBP2	AE binding protein 2	121536	yes			yes		2	
AFF4	AF4/FMR2 family, member 4	27125	yes		yes			2	
AGBL5	ATP/GTP binding protein-like 5	60509	yes				yes	2	
AGPS	alkylglycerone phosphate synthase	8540	yes			yes		2	
AHCYL2	adenosylhomocysteinase-like 2	23382	yes			yes		2	
AICDA	activation-induced cytidine deaminase	57379	yes			yes		2	
AK3	adenylate kinase 3	50808	yes			yes		2	
AK5	adenylate kinase 5	26289	yes			yes		2	
AKAP5	A kinase (PRKA) anchor protein 5	9495	yes			yes		2	
AKNAD1	chromosome 1 open reading frame 62	254268	yes				yes	2	
ALDH6A1	aldehyde dehydrogenase 6 family, member A1	4329	yes				yes	2	
ALS2CR8	amyotrophic lateral sclerosis 2 (juvenile) chromosome	79800	yes				yes	2	
ANKLE2	ankyrin repeat and LEM domain containing 2	23141	yes			yes		2	
ANKRD33	ankyrin repeat domain 33	341405	yes				yes	2	
ANKRD46	ankyrin repeat domain 46	157567	yes			yes		2	
ANKRD57	ankyrin repeat domain 57	65124	yes			yes		2	
ANO6	anoctamin 6	196527	yes			yes		2	
AP1S3	adaptor-related protein complex 1, sigma 3 subunit	130340	yes				yes	2	
APOL6	apolipoprotein L, 6	80830			yes	yes		2	
ARF4	ADP-ribosylation factor 4	378	yes				yes	2	
ARPC5L	actin related protein 2/3 complex, subunit 5-like	81873	yes				yes	2	
ARRDC3	arrestin domain containing 3	57561	yes			yes		2	
ART1	ADP-ribosyltransferase 1	417	yes				yes	2	
ASCC3	activating signal cointegrator 1 complex subunit 3	10973	yes	yes				2	
ASTN1	astrotactin 1	460	yes				yes	2	
ATAD1	ATPase family, AAA domain containing 1	84896	yes			yes		2	
ATP13A3	ATPase type 13A3	79572	yes				yes	2	
ATP1B4	ATPase, (Na+)/K+ transporting, beta 4 polypeptide	23439	yes			yes		2	
ATP6V0A2	ATPase, H+ transporting, lysosomal V0 subunit a2	23545	yes				yes	2	
ATP6V1E1	ATPase, H+ transporting, lysosomal 31kDa, V1	529	yes				yes	2	
ATP6V1E2	ATPase, H+ transporting, lysosomal 31kDa, V1	90423	yes				yes	2	
B9D1	B9 protein domain 1	27077	yes				yes	2	
BCCIP	BRCA2 and CDKN1A interacting protein	56647	yes				yes	2	
BCDIN3D	BCDIN3 domain containing	144233	yes				yes	2	
BDNF	brain-derived neurotrophic factor	627	yes	yes				2	
BHMT	betaine-homocysteine methyltransferase	635	yes			yes		2	
BLMH	bleomycin hydrolase	642	yes				yes	2	
BMP2K	BMP2 inducible kinase	55589	yes				yes	2	
BOLL	bol, boule-like (Drosophila)	66037	yes				yes	2	
BRCC3	BRCA1/BRCA2-containing complex, subunit 3	79184	yes			yes		2	
BRIP1	BRCA1 interacting protein C-terminal helicase 1	83990	yes				yes	2	
C10ORF122	chromosome 10 open reading frame 122	387718	yes				yes	2	
C10ORF72	chromosome 10 open reading frame 72	196740	yes			yes		2	
C10ORF84	chromosome 10 open reading frame 84	63877	yes				yes	2	
C11ORF41	chromosome 11 open reading frame 41	25758	yes	yes				2	
C13ORF23	chromosome 13 open reading frame 23	80209	yes			yes		2	
C13ORF31	chromosome 13 open reading frame 31	144811	yes				yes	2	
C16ORF46	chromosome 16 open reading frame 46	123775	yes				yes	2	
C17ORF78	chromosome 17 open reading frame 78	284099	yes				yes	2	
C19ORF55	chromosome 19 open reading frame 55	148137	yes			yes		2	
C1ORF65	chromosome 1 open reading frame 65	164127	yes				yes	2	
C1ORF94	chromosome 1 open reading frame 94	84970	yes				yes	2	
C20ORF112	chromosome 20 open reading frame 112	140688	yes				yes	2	
C20ORF194	chromosome 20 open reading frame 194	25943	yes				yes	2	
C22ORF25	chromosome 22 open reading frame 25	128989	yes				yes	2	
C2ORF3	chromosome 2 open reading frame 3	6936	yes	yes				2	
C2ORF67	chromosome 2 open reading frame 67	151050	yes	yes				2	
C3ORF18	chromosome 3 open reading frame 18	51161	yes				yes	2	
C3ORF24	chromosome 3 open reading frame 24	115795	yes				yes	2	
C3ORF37	chromosome 3 open reading frame 37	56941	yes				yes	2	
C3ORF55	chromosome 3 open reading frame 55	152078	yes			yes		2	
C3ORF59	chromosome 3 open reading frame 59	151963	yes			yes		2	
C5ORF22	chromosome 5 open reading frame 22	55322	yes			yes		2	
C5ORF33	chromosome 5 open reading frame 33	133668	yes				yes	2	
C6ORF153	chromosome 6 open reading frame 153	88745	yes				yes	2	
C7ORF53	chromosome 7 open reading frame 53	286006	yes			yes		2	
C8ORF76	chromosome 8 open reading frame 76	84933	yes				yes	2	
CAB39L	calcium binding protein 39-like	81617	yes			yes		2	
CACNG1	calcium channel, voltage-dependent, gamma subunit	786	yes				yes	2	
CALCRL	calcitonin receptor-like	10203	yes	yes				2	
CALHM2	calcium homeostasis modulator 2	51063	yes				yes	2	
CARD8	caspase recruitment domain family, member 8	22900	yes		yes			2	

Appendix

Symbol	Gene name	Entrez	MR	TS	µT	DB	MC	#	KEGG
CBLB	Cas-Br-M (murine) ecotropic retroviral transforming	868	yes				yes	2	
CC2D2B	coiled-coil and C2 domain containing 2B	387707	yes				yes	2	
CCDC152	coiled-coil domain containing 152	100129	yes			yes		2	
CCDC59	coiled-coil domain containing 59	29080	yes			yes		2	
CCDC90B	coiled-coil domain containing 90B	60492	yes				yes	2	
CCDC91	coiled-coil domain containing 91	55297	yes			yes		2	
CCNDBP1	cyclin D-type binding-protein 1	23582	yes				yes	2	
CDADC1	cytidine and dCMP deaminase domain containing 1	81602	yes				yes	2	
CDC123	cell division cycle 123 homolog (S. cerevisiae)	8872	yes				yes	2	
CDC14A	CDC14 cell division cycle 14 homolog A (S.	8556	yes			yes		2	
CDHR3	hypothetical protein FLJ23834	222256	yes				yes	2	
CDK7	cyclin-dependent kinase 7	1022	yes				yes	2	
CENPC1	centromere protein C 1	1060	yes				yes	2	
CEP78	centrosomal protein 78kDa	84131	yes				yes	2	
CFHR3	complement factor H-related 3	10878	yes				yes	2	
CFHR4	complement factor H-related 4	10877	yes				yes	2	
CHIC2	cysteine-rich hydrophobic domain 2	26511	yes				yes	2	
CLDN8	claudin 8	9073	yes	yes				2	
CLEC1A	C-type lectin domain family 1, member A	51267	yes				yes	2	
CLEC2D	C-type lectin domain family 2, member D	29121	yes				yes	2	
CLEC3A	C-type lectin domain family 3, member A	10143	yes				yes	2	
CLIC2	chloride intracellular channel 2	1193	yes			yes		2	
CLPX	ClpX caseinolytic peptidase X homolog (E. coli)	10845	yes				yes	2	
CMTM8	CKLF-like MARVEL transmembrane domain	152189	yes				yes	2	
CNGB3	cyclic nucleotide gated channel beta 3	54714	yes	yes				2	
CNOT8	CCR4-NOT transcription complex, subunit 8	9337	yes			yes		2	
CNTN3	contactin 3 (plasmacytoma associated)	5067	yes				yes	2	
COCH	coagulation factor C homolog, cochlin (Limulus	1690	yes				yes	2	
COL14A1	collagen, type XIV, alpha 1	7373	yes				yes	2	
COL4A4	collagen, type IV, alpha 4	1286	yes			yes		2	
COMMD8	COMM domain containing 8	54951	yes			yes		2	
CPT1A	carnitine palmitoyltransferase 1A (liver)	1374	yes				yes	2	
CR1	complement component (3b/4b) receptor 1 (Knops	1378	yes		yes			2	
CRB1	crumbs homolog 1 (Drosophila)	23418	yes		yes			2	
CRBN	cereblon	51185	yes				yes	2	
CREBBP	CREB binding protein	1387	yes	yes				2	
CREG1	cellular repressor of E1A-stimulated genes 1	8804	yes			yes		2	
CRISPLD1	cysteine-rich secretory protein LCCL domain	83690	yes			yes		2	
CSTA	cystatin A (stefin A)	1475	yes				yes	2	
CTNND1	catenin (cadherin-associated protein), delta 1	1500	yes	yes				2	
CTNND2	catenin (cadherin-associated protein), delta 2 (neural	1501	yes	yes				2	
CTPS2	CTP synthase II	56474	yes				yes	2	
CYB5D1	cytochrome b5 domain containing 1	124637	yes			yes		2	
CYP2U1	cytochrome P450, family 2, subfamily U, polypeptide	113612	yes				yes	2	
CYP3A4	cytochrome P450, family 3, subfamily A, polypeptide	1576	yes				yes	2	
CYP3A7	cytochrome P450, family 3, subfamily A, polypeptide	1551	yes				yes	2	
CYP7B1	cytochrome P450, family 7, subfamily B, polypeptide	9420	yes				yes	2	
CYYR1	cysteine/tyrosine-rich 1	116159	yes			yes		2	
DDX20	DEAD (Asp-Glu-Ala-Asp) box polypeptide 20	11218	yes				yes	2	
DDX52	DEAD (Asp-Glu-Ala-Asp) box polypeptide 52	11056	yes	yes				2	
DEM1	defects in morphology 1 homolog (S. cerevisiae)	64789	yes				yes	2	
DPY19L4	dpy-19-like 4 (C. elegans)	286148	yes			yes		2	
DSEL	dermatan sulfate epimerase-like	92126	yes			yes		2	
DUS4L	dihydrouridine synthase 4-like (S. cerevisiae)	11062	yes				yes	2	
DUSP6	dual specificity phosphatase 6	1848	yes			yes		2	
EDIL3	EGF-like repeats and discoidin I-like domains 3	10085	yes			yes		2	
EFHA2	EF-hand domain family, member A2	286097	yes			yes		2	
EFHC2	EF-hand domain (C-terminal) containing 2	80258	yes			yes		2	
EIF1AX	eukaryotic translation initiation factor 1A, X-linked	1964	yes			yes		2	
EIF5B	eukaryotic translation initiation factor 5B	9669	yes			yes		2	
ENDOD1	endonuclease domain containing 1	23052	yes			yes		2	
ENPP4	ectonucleotide pyrophosphatase/phosphodiesterase 4	22875	yes			yes		2	
EPB41	erythrocyte membrane protein band 4.1 (elliptocytosis	2035	yes	yes				2	
EPS8L3	EPS8-like 3	79574	yes				yes	2	
ERC2	ELKS/RAB6-interacting/CAST family member 2	26059	yes			yes		2	
ERI2	exoribonuclease 2	112479	yes				yes	2	
ERMN	ermin, ERM-like protein	57471	yes			yes		2	
EXO1	exonuclease 1	9156	yes				yes	2	
EYS	eyes shut homolog (Drosophila)	346007	yes				yes	2	
FADS2	fatty acid desaturase 2	9415	yes				yes	2	
FAM13C	family with sequence similarity 13, member C	220965	yes				yes	2	
FAM162B	family with sequence similarity 162, member B	221303	yes				yes	2	
FAM188A	chromosome 10 open reading frame 97	80013	yes				yes	2	
FAM26D	family with sequence similarity 26, member D	221301	yes				yes	2	
FAM36A	family with sequence similarity 36, member A	116228	yes			yes		2	
FAM63B	family with sequence similarity 63, member B	54629	yes			yes		2	
FAM96A	family with sequence similarity 96, member A	84191	yes				yes	2	
FAS	Fas (TNF receptor superfamily, member 6)	355	yes		yes			2	
FGF1	fibroblast growth factor 1 (acidic)	2246	yes		yes			2	
FGF5	fibroblast growth factor 5	2250	yes		yes			2	
FKBP14	FK506 binding protein 14, 22 kDa	55033	yes				yes	2	
FKBP15	FK506 binding protein 15, 133kDa	23307	yes			yes		2	

Appendix

Symbol	Gene name	Entrez	MR	TS	µT	DB	MC	#	KEGG
FKBP4	FK506 binding protein 4, 59kDa	2288	yes				yes	2	
FKBP9L	FK506 binding protein 9-like	360132	yes				yes	2	
FLRT2	fibronectin leucine rich transmembrane protein 2	23768	yes			yes		2	
FMO5	flavin containing monooxygenase 5	2330	yes				yes	2	
FNBP1	formin binding protein 1	23048	yes				yes	2	
FOLR1	folate receptor 1 (adult)	2348	yes				yes	2	
FOXM1	forkhead box M1	2305	yes	yes				2	
FOXN2	forkhead box N2	3344	yes			yes		2	
FRYL	FRY-like	285527	yes			yes		2	
FUCA1	fucosidase, alpha-L- 1, tissue	2517	yes				yes	2	
FUNDC2	FUN14 domain containing 2	65991	yes			yes		2	
FUT5	fucosyltransferase 5 (alpha (1,3) fucosyltransferase)	2527	yes				yes	2	
FUT9	fucosyltransferase 9 (alpha (1,3) fucosyltransferase)	10690		yes		yes		2	
FZD3	frizzled homolog 3 (Drosophila)	7976	yes			yes		2	
GAN	gigaxonin	8139	yes				yes	2	
GAPT	GRB2-binding adaptor protein, transmembrane	202309	yes				yes	2	
GARNL3	GTPase activating Rap/RanGAP domain-like 3	84253	yes				yes	2	
GBE1	glucan (1,4-alpha-), branching enzyme 1	2632	yes				yes	2	
GCET2	germinal center expressed transcript 2	257144	yes			yes		2	
GCNT1	glucosaminyl (N-acetyl) transferase 1, core 2 (beta-	2650	yes			yes		2	
GEN1	Gen homolog 1, endonuclease (Drosophila)	348654	yes			yes		2	
GIGYF2	GRB10 interacting GYF protein 2	26058	yes	yes				2	
GLI1	GLI family zinc finger 1	2735	yes				yes	2	
GLIPR1	GLI pathogenesis-related 1	11010	yes			yes		2	
GLOD4	glyoxalase domain containing 4	51031	yes				yes	2	
GLRX	glutaredoxin (thioltransferase)	2745	yes				yes	2	
GNPAT	glyceronephosphate O-acyltransferase	8443	yes				yes	2	
GOLGA8E	golgi autoantigen, golgin subfamily a, 8E	390535	yes			yes		2	
GPLD1	glycosylphosphatidylinositol specific phospholipase	2822	yes				yes	2	
GPR85	G protein-coupled receptor 85	54329	yes		yes			2	
GRHPR	glyoxylate reductase/hydroxypyruvate reductase	9380	yes				yes	2	
GRIP1	glutamate receptor interacting protein 1	23426	yes				yes	2	
GSG1	germ cell associated 1	83445	yes				yes	2	
GSG2	germ cell associated 2 (haspin)	83903	yes				yes	2	
GSR	glutathione reductase	2936	yes				yes	2	
GSTA2	glutathione S-transferase alpha 2	2939	yes				yes	2	
GTF2E1	general transcription factor IIE, polypeptide 1, alpha	2960	yes	yes				2	
GUF1	GUF1 GTPase homolog (S. cerevisiae)	60558	yes				yes	2	
GYG1	glycogenin 1	2992	yes				yes	2	
HAT1	histone acetyltransferase 1	8520	yes				yes	2	
HDAC9	histone deacetylase 9	9734	yes			yes		2	
HECTD2	HECT domain containing 2	143279	yes			yes		2	
HERC4	hect domain and RLD 4	26091	yes		yes			2	
HFE	hemochromatosis	3077	yes			yes		2	
HMGB2	high-mobility group box 2	3148	yes				yes	2	
HMGN4	high mobility group nucleosomal binding domain 4	10473	yes			yes		2	
HNRNPA2B	heterogeneous nuclear ribonucleoprotein A2/B1	3181	yes			yes		2	
HOMER1	homer homolog 1 (Drosophila)	9456	yes			yes		2	
HOOK1	hook homolog 1 (Drosophila)	51361	yes			yes		2	
HS2ST1	heparan sulfate 2-O-sulfotransferase 1	9653	yes			yes		2	
ICA1L	islet cell autoantigen 1,69kDa-like	130026	yes				yes	2	
IFT80	intraflagellar transport 80 homolog (Chlamydomonas)	57560	yes			yes		2	
IL13RA1	interleukin 13 receptor, alpha 1	3597	yes			yes		2	
INO80C	INO80 complex subunit C	125476	yes				yes	2	
ITGAV	integrin, alpha V (vitronectin receptor, alpha	3685	yes			yes		2	
JKAMP	chromosome 14 open reading frame 100	51528	yes				yes	2	
KCMF1	potassium channel modulatory factor 1	56888	yes	yes				2	
KCNA2	potassium voltage-gated channel, shaker-related	3737	yes				yes	2	
KCNA5	potassium voltage-gated channel, shaker-related	3741	yes			yes		2	
KCNJ3	potassium inwardly-rectifying channel, subfamily J,	3760	yes			yes		2	
KCTD6	potassium channel tetramerisation domain containing	200845	yes			yes		2	
KDELC2	KDEL (Lys-Asp-Glu-Leu) containing 2	143888	yes	yes				2	
KDELR1	KDEL (Lys-Asp-Glu-Leu) endoplasmic reticulum	10945	yes				yes	2	
KIAA1383	KIAA1383	54027	yes			yes		2	
KIAA1409	KIAA1409	57578	yes				yes	2	
KLHDC5	kelch domain containing 5	57542	yes			yes		2	
KLHL7	kelch-like 7 (Drosophila)	55975	yes			yes		2	
KLRG1	killer cell lectin-like receptor subfamily G, member 1	10219	yes			yes		2	
KPNB1	karyopherin (importin) beta 1	3837	yes	yes				2	
KRT32	keratin 32	3882	yes				yes	2	
KRTAP1-5	keratin associated protein 1-5	83895	yes			yes		2	
LACTB	lactamase, beta	114294	yes			yes		2	
LCOR	ligand dependent nuclear receptor corepressor	84458	yes			yes		2	
LGR4	leucine-rich repeat-containing G protein-coupled	55366	yes			yes		2	
LGSN	lengsin, lens protein with glutamine synthetase	51557	yes				yes	2	
LIX1	Lix1 homolog (chicken)	167410	yes				yes	2	
LMOD3	leiomodin 3 (fetal)	56203	yes			yes		2	
LMX1A	LIM homeobox transcription factor 1, alpha	4009	yes			yes		2	
LNX2	ligand of numb-protein X 2	222484	yes			yes		2	
LPCAT1	lysophosphatidylcholine acyltransferase 1	79888	yes	yes				2	
LPPR5	phosphatidic acid phosphatase type 2	163404	yes		yes			2	
LRRC27	leucine rich repeat containing 27	80313			yes	yes		2	

Appendix

Symbol	Gene name	Entrez	MR	TS	µT	DB	MC	#	KEGG
LRRIQ3	leucine-rich repeats and IQ motif containing 3	127255	yes				yes	2	
LUC7L	LUC7-like (S. cerevisiae)	55692	yes				yes	2	
MAGEE1	melanoma antigen family E, 1	57692	yes				yes	2	
MAK16	MAK16 homolog (S. cerevisiae)	84549	yes			yes		2	
MCHR2	melanin-concentrating hormone receptor 2	84539	yes			yes		2	
MCL1	myeloid cell leukemia sequence 1 (BCL2-related)	4170	yes			yes		2	
MEAF6	chromosome 1 open reading frame 149	64769	yes				yes	2	
MEOX2	mesenchyme homeobox 2	4223	yes				yes	2	
MERTK	c-mer proto-oncogene tyrosine kinase	10461	yes				yes	2	
MEST	mesoderm specific transcript homolog (mouse)	4232	yes				yes	2	
MEX3C	mex-3 homolog C (C. elegans)	51320	yes			yes		2	
MFAP3	microfibrillar-associated protein 3	4238	yes			yes		2	
MGAT4A	mannosyl (alpha-1,3-)-glycoprotein beta-1,4-N-	11320		yes	yes			2	
MID1	midline 1 (Opitz/BBB syndrome)	4281	yes			yes		2	
MIF4GD	MIF4G domain containing	57409	yes			yes		2	
MLF1	myeloid leukemia factor 1	4291	yes				yes	2	
MLH1	mutL homolog 1, colon cancer, nonpolyposis type 2	4292	yes				yes	2	
MLLT6	myeloid/lymphoid or mixed-lineage leukemia (trithorax	4302	yes	yes				2	
MMD	monocyte to macrophage differentiation-associated	23531	yes			yes		2	
MMP16	matrix metallopeptidase 16 (membrane-inserted)	4325	yes			yes		2	
MMP21	matrix metallopeptidase 21	118856	yes				yes	2	
MRPL13	mitochondrial ribosomal protein L13	28998	yes				yes	2	
MRS2	MRS2 magnesium homeostasis factor homolog (S.	57380	yes			yes		2	
MTMR10	myotubularin related protein 10	54893	yes			yes		2	
MTMR4	myotubularin related protein 4	9110	yes			yes		2	
MTX3	metaxin 3	345778	yes			yes		2	
MYCBP	c-myc binding protein	26292	yes	yes				2	
MYH9	myosin, heavy chain 9, non-muscle	4627	yes				yes	2	
MYO16	myosin XVI	23026	yes				yes	2	
MYT1L	myelin transcription factor 1-like	23040	yes	yes				2	
NAALADL2	N-acetylated alpha-linked acidic dipeptidase-like 2	254827	yes			yes		2	
NANP	N-acetylneuraminic acid phosphatase	140838	yes			yes		2	
NDRG3	NDRG family member 3	57446	yes				yes	2	
NEDD4	neural precursor cell expressed, developmentally	4734	yes			yes		2	
NFATC3	nuclear factor of activated T-cells, cytoplasmic,	4775	yes				yes	2	yes
NHS	Nance-Horan syndrome (congenital cataracts and	4810	yes	yes				2	
NOL8	nucleolar protein 8	55035	yes				yes	2	
NR4A2	nuclear receptor subfamily 4, group A, member 2	4929	yes			yes		2	
NUP153	nucleoporin 153kDa	9972	yes			yes		2	
NUP62CL	nucleoporin 62kDa C-terminal like	54830	yes				yes	2	
NXPH2	neurexophilin 2	11249	yes				yes	2	
NXT1	NTF2-like export factor 1	29107	yes				yes	2	
ODF2L	outer dense fiber of sperm tails 2-like	57489	yes		yes			2	
OGFR	opioid growth factor receptor	11054	yes				yes	2	
ONECUT2	one cut homeobox 2	9480			yes	yes		2	
OPTN	optineurin	10133	yes			yes		2	
OR51E1	olfactory receptor, family 51, subfamily E, member 1	143503	yes				yes	2	
OSBPL11	oxysterol binding protein-like 11	114885	yes			yes		2	
P4HA1	prolyl 4-hydroxylase, alpha polypeptide I	5033	yes				yes	2	
P4HA2	prolyl 4-hydroxylase, alpha polypeptide II	8974	yes				yes	2	
PANX2	pannexin 2	56666	yes	yes				2	
PCDH15	protocadherin 15	65217	yes				yes	2	
PCDH9	protocadherin 9	5101	yes			yes		2	
PCDHB4	protocadherin beta 4	56131	yes			yes		2	
PCGF5	polycomb group ring finger 5	84333	yes		yes			2	
PCLO	piccolo (presynaptic cytomatrix protein)	27445	yes			yes		2	
PCMT1	protein-L-isoaspartate (D-aspartate) O-	5110	yes			yes		2	
PCYOX1	prenylcysteine oxidase 1	51449	yes			yes		2	
PDE10A	phosphodiesterase 10A	10846	yes		yes			2	
PDE3A	phosphodiesterase 3A, cGMP-inhibited	5139	yes				yes	2	
PDE4D	phosphodiesterase 4D, cAMP-specific	5144	yes			yes		2	
PDIK1L	PDLIM1 interacting kinase 1 like	149420	yes			yes		2	
PDK4	pyruvate dehydrogenase kinase, isozyme 4	5166	yes	yes				2	
PEPD	peptidase D	5184	yes				yes	2	
PER3	period homolog 3 (Drosophila)	8863	yes			yes		2	
PERP	PERP, TP53 apoptosis effector	64065	yes			yes		2	
PGAP1	post-GPI attachment to proteins 1	80055			yes	yes		2	
PGBD2	piggyBac transposable element derived 2	267002	yes				yes	2	
PHLDA1	pleckstrin homology-like domain, family A, member 1	22822	yes			yes		2	
PHYH	phytanoyl-CoA 2-hydroxylase	5264	yes				yes	2	
PIK3R1	phosphoinositide-3-kinase, regulatory subunit 1	5295	yes	yes				2	yes
PITPNC1	phosphatidylinositol transfer protein, cytoplasmic 1	26207	yes			yes		2	
PLA2G4A	phospholipase A2, group IVA (cytosolic, calcium-	5321	yes				yes	2	yes
PLAU	plasminogen activator, urokinase	5328	yes				yes	2	
PLD1	phospholipase D1, phosphatidylcholine-specific	5337	yes			yes		2	
PLEKHA2	pleckstrin homology domain containing, family A	59339	yes			yes		2	
PLP1	proteolipid protein 1	5354	yes			yes		2	
PMP2	peripheral myelin protein 2	5375	yes			yes		2	
POF1B	premature ovarian failure, 1B	79983	yes			yes		2	
POLE2	polymerase (DNA directed), epsilon 2 (p59 subunit)	5427	yes				yes	2	
POLH	polymerase (DNA directed), eta	5429	yes			yes		2	
POLR2K	polymerase (RNA) II (DNA directed) polypeptide K,	5440	yes				yes	2	

Appendix

Symbol	Gene name	Entrez	MR	TS	µT	DB	MC	#	KEGG
PROCR	protein C receptor, endothelial (EPCR)	10544	yes				yes	2	
PROL1	proline rich, lacrimal 1	58503	yes				yes	2	
PRSS12	protease, serine, 12 (neurotrypsin, motopsin)	8492	yes			yes		2	
PRTG	protogenin homolog (Gallus gallus)	283659	yes				yes	2	
PTH	parathyroid hormone	5741	yes				yes	2	
RAB11FIP2	RAB11 family interacting protein 2 (class I)	22841	yes			yes		2	
RAB12	RAB12, member RAS oncogene family	201475	yes			yes		2	
RAB4A	RAB4A, member RAS oncogene family	5867	yes				yes	2	
RAD18	RAD18 homolog (S. cerevisiae)	56852	yes				yes	2	
RAD23A	RAD23 homolog A (S. cerevisiae)	5886	yes	yes				2	
RAD23B	RAD23 homolog B (S. cerevisiae)	5887	yes	yes				2	
RAG1	recombination activating gene 1	5896	yes			yes		2	
RAP1GDS1	RAP1, GTP-GDP dissociation stimulator 1	5910	yes				yes	2	
RB1	retinoblastoma 1	5925	yes			yes		2	
RBBP6	retinoblastoma binding protein 6	5930	yes				yes	2	
RBM27	RNA binding motif protein 27	54439	yes			yes		2	
RCN2	reticulocalbin 2, EF-hand calcium binding domain	5955	yes			yes		2	
REV1	REV1 homolog (S. cerevisiae)	51455	yes				yes	2	
RHOA	ras homolog gene family, member A	387	yes			yes		2	
RIF1	RAP1 interacting factor homolog (yeast)	55183	yes				yes	2	
RIN1	Ras and Rab Interactor 1	9610	yes				yes	2	
RNF11	ring finger protein 11	26994	yes			yes		2	
RNF144A	ring finger protein 144A	9781	yes				yes	2	
RNF8	ring finger protein 8	9025	yes				yes	2	
RNGTT	RNA guanylyltransferase and 5'-phosphatase	8732	yes					2	
ROM1	retinal outer segment membrane protein 1	6094	yes				yes	2	
RORB	RAR-related orphan receptor B	6096	yes			yes		2	
RPA1	replication protein A1, 70kDa	6117	yes	yes				2	
RPGRIP1L	RPGRIP1-like	23322	yes				yes	2	
RPL34	ribosomal protein L34	6164	yes				yes	2	
RPN2	ribophorin II	6185	yes				yes	2	
RQCD1	RCD1 required for cell differentiation1 homolog (S.	9125	yes				yes	2	
RTN3	reticulon 3	10313	yes			yes		2	
SAMD3	sterile alpha motif domain containing 3	154075	yes				yes	2	
SC5DL	sterol-C5-desaturase (ERG3 delta-5-desaturase	6309	yes	yes				2	
SDCBP	syndecan binding protein (syntenin)	6386	yes			yes		2	
SFRS12	splicing factor, arginine/serine-rich 12	140890	yes	yes				2	
SFRS5	splicing factor, arginine/serine-rich 5	6430	yes				yes	2	
SFXN3	sideroflexin 3	81855	yes				yes	2	
SHISA2	shisa homolog 2 (Xenopus laevis)	387914	yes			yes		2	
SIRT5	sirtuin (silent mating type information regulation 2	23408	yes			yes		2	
SLAMF1	signaling lymphocytic activation molecule family	6504	yes			yes		2	
SLC12A2	solute carrier family 12 (sodium/potassium/chloride	6558	yes			yes		2	
SLC16A9	solute carrier family 16, member 9 (monocarboxylic	220963	yes				yes	2	
SLC22A7	solute carrier family 22 (organic anion transporter),	10864	yes				yes	2	
SLC23A2	solute carrier family 23 (nucleobase transporters),	9962	yes				yes	2	
SLC26A7	solute carrier family 26, member 7	115111	yes				yes	2	
SLC36A3	solute carrier family 36 (proton/amino acid symporter),	285641	yes				yes	2	
SLC41A1	solute carrier family 41, member 1	254428	yes	yes				2	
SLC5A3	solute carrier family 5 (sodium/myo-inositol	6526			yes	yes		2	
SLC6A4	solute carrier family 6 (neurotransmitter transporter,	6532	yes				yes	2	
SLC7A6OS	solute carrier family 7, member 6 opposite strand	84138	yes	yes				2	
SMAD2	SMAD family member 2	4087			yes	yes		2	
SMPD3	sphingomyelin phosphodiesterase 3, neutral	55512	yes	yes				2	
SND1	staphylococcal nuclease and tudor domain containing	27044	yes				yes	2	
SNRNP27	small nuclear ribonucleoprotein 27kDa (U4/U6.U5)	11017	yes				yes	2	
SOHLH2	spermatogenesis and oogenesis specific basic helix-	54937	yes	yes				2	
SON	SON DNA binding protein	6651	yes			yes		2	
SPAG16	sperm associated antigen 16	79582	yes				yes	2	
SPP2	secreted phosphoprotein 2, 24kDa	6694	yes				yes	2	
SPRR2F	small proline-rich protein 2F	6705	yes				yes	2	
ST8SIA1	ST8 alpha-N-acetyl-neuraminide alpha-2,8-	6489		yes		yes		2	
STAU1	staufen, RNA binding protein, homolog 1 (Drosophila)	6780	yes				yes	2	
STK32A	serine/threonine kinase 32A	202374	yes			yes		2	
SUV420H1	suppressor of variegation 4-20 homolog 1	51111	yes				yes	2	
SV2A	synaptic vesicle glycoprotein 2A	9900	yes				yes	2	
SYF2	SYF2 homolog, RNA splicing factor (S. cerevisiae)	25949	yes				yes	2	
SYT10	synaptotagmin X	341359	yes			yes		2	
SYT15	synaptotagmin XV	83849	yes				yes	2	
TAF11	TAF11 RNA polymerase II, TATA box binding protein	6882	yes				yes	2	
TAF13	TAF13 RNA polymerase II, TATA box binding protein	6884	yes				yes	2	
TALDO1	transaldolase 1	6888	yes				yes	2	
TBC1D24	TBC1 domain family, member 24	57465	yes			yes		2	
TCEAL5	transcription elongation factor A (SII)-like 5	340543	yes				yes	2	
TCEAL6	transcription elongation factor A (SII)-like 6	158931	yes				yes	2	
TET1	tet oncogene 1	80312	yes			yes		2	
TET3	tet oncogene family member 3	200424		yes			yes	2	
TEX15	testis expressed 15	56154	yes			yes		2	
TFDP3	transcription factor Dp family, member 3	51270	yes				yes	2	
TGDS	TDP-glucose 4,6-dehydratase	23483	yes				yes	2	
TGFBR1	transforming growth factor, beta receptor 1	7046	yes	yes				2	
TIPIN	TIMELESS interacting protein	54962	yes			yes		2	

Symbol	Gene name	Entrez	MR	TS	µT	DB	MC	#	KEGG
TIPRL	TIP41, TOR signaling pathway regulator-like (S.	261726	yes				yes	2	
TKTL1	transketolase-like 1	8277	yes			yes		2	
TKTL2	transketolase-like 2	84076	yes			yes		2	
TM2D1	TM2 domain containing 1	83941	yes				yes	2	
TM4SF20	transmembrane 4 L six family member 20	79853	yes				yes	2	
TMEM14E	transmembrane protein 14E	645843	yes			yes		2	
TMEM174	transmembrane protein 174	134288	yes				yes	2	
TMEM185B	transmembrane protein 185B (pseudogene)	79134	yes				yes	2	
TMEM215	transmembrane protein 215	401498	yes			yes		2	
TMEM232	hypothetical protein LOC642987	642987	yes				yes	2	
TNPO1	transportin 1	3842	yes		yes			2	
TRAF3	TNF receptor-associated factor 3	7187	yes				yes	2	
TRIM58	tripartite motif-containing 58	25893	yes			yes		2	
TRIM7	tripartite motif-containing 7	81786	yes				yes	2	
TSGA10	testis specific, 10	80705	yes			yes		2	
TTC30A	tetratricopeptide repeat domain 30A	92104	yes			yes		2	
TWIST1	twist homolog 1 (Drosophila)	7291	yes				yes	2	
TWSG1	twisted gastrulation homolog 1 (Drosophila)	57045	yes			yes		2	
UBE2H	ubiquitin-conjugating enzyme E2H (UBC8 homolog,	7328	yes	yes				2	
UCHL3	ubiquitin carboxyl-terminal esterase L3 (ubiquitin	7347	yes				yes	2	
UGT2A3	UDP glucuronosyltransferase 2 family, polypeptide A3	79799	yes			yes		2	
UGT8	UDP glycosyltransferase 8	7368	yes			yes		2	
UHRF2	ubiquitin-like with PHD and ring finger domains 2	115426	yes				yes	2	
USP42	ubiquitin specific peptidase 42	84132	yes				yes	2	
USP47	ubiquitin specific peptidase 47	55031	yes				yes	2	
USP6	ubiquitin specific peptidase 6 (Tre-2 oncogene)	9098	yes			yes		2	
UTP14A	UTP14, U3 small nucleolar ribonucleoprotein,	10813	yes				yes	2	
VAMP3	vesicle-associated membrane protein 3 (cellubrevin)	9341	yes				yes	2	
VEZF1	vascular endothelial zinc finger 1	7716	yes	yes				2	
VPS13A	vacuolar protein sorting 13 homolog A (S. cerevisiae)	23230	yes				yes	2	
VPS13B	vacuolar protein sorting 13 homolog B (yeast)	157680	yes		yes			2	
VSX1	visual system homeobox 1	30813	yes				yes	2	
VWDE	von Willebrand factor D and EGF domains	221806	yes				yes	2	
WARS2	tryptophanyl tRNA synthetase 2, mitochondrial	10352	yes				yes	2	
WDR26	WD repeat domain 26	80232	yes			yes		2	
WDR67	WD repeat domain 67	93594	yes				yes	2	
WT1	Wilms tumor 1	7490	yes			yes		2	
XAF1	XIAP associated factor 1	54739	yes			yes		2	
YAP1	Yes-associated protein 1, 65kDa	10413	yes			yes		2	
YIPF4	Yip1 domain family, member 4	84272	yes			yes		2	
YWHAB	tyrosine 3-monooxygenase/tryptophan 5-	7529	yes			yes		2	
YWHAH	tyrosine 3-monooxygenase/tryptophan 5-	7533	yes				yes	2	
ZBTB10	zinc finger and BTB domain containing 10	65986	yes			yes		2	
ZBTB26	zinc finger and BTB domain containing 26	57684	yes				yes	2	
ZBTB38	zinc finger and BTB domain containing 38	253461	yes	yes				2	
ZCCHC11	zinc finger, CCHC domain containing 11	23318	yes				yes	2	
ZFPM2	zinc finger protein, multitype 2	23414	yes	yes				2	
ZFR	zinc finger RNA binding protein	51663	yes				yes	2	
ZHX3	zinc fingers and homeoboxes 3	23051	yes	yes				2	
ZNF160	zinc finger protein 160	90338	yes				yes	2	
ZNF254	zinc finger protein 254	9534	yes				yes	2	
ZNF271	zinc finger protein 271	10778	yes			yes		2	
ZNF396	zinc finger protein 396	252884	yes			yes		2	
ZNF434	zinc finger protein 434	54925	yes			yes		2	
ZNF449	zinc finger protein 449	203523	yes				yes	2	
ZNF460	zinc finger protein 460	10794	yes			yes		2	
ZNF470	zinc finger protein 470	388566	yes			yes		2	
ZNF490	zinc finger protein 490	57474	yes			yes		2	
ZNF510	zinc finger protein 510	22869	yes			yes		2	
ZNF516	zinc finger protein 516	9658	yes	yes				2	
ZNF521	zinc finger protein 521	25925	yes	yes				2	
ZNF599	zinc finger protein 599	148103	yes				yes	2	
ZRANB1	zinc finger, RAN-binding domain containing 1	54764	yes	yes				2	
ZXDB	zinc finger, X-linked, duplicated B	158586	yes	yes				2	

7.6 Gene set enrichment analysis of microRNA 361-5p, Pum1 and Pum2 targets

MicroRNA target predictions were from the following web services: microRNA.org (Betel *et al.*, 2010), TargetScan (Friedman *et al.*, 2009), DIANA-microT (Maragkakis *et al.*, 2009), miRDB (Wang, 2008), and MicroCosm (Griffiths-Jones *et al.*, 2008). Experimentally verified Pum1 and Pum2 targets were previously published (Galgano *et al.*, 2008; Morris *et al.*, 2008; Hafner *et al.*, 2010b). Results were pooled and converted to Entrez gene identifiers using the DAVID web service (Huang *et al.*, 2008). Putative targets were compared to a human reference gene list and analyzed for pathway enrichment using PANTHER (Thomas *et al.*, 2006).

Appendix

Pathways significantly enriched among predicted microRNA 361-5p (miR-361-5p) or experimentally verified Pum1 or Pum2 targets are listed together with their P values.

Pathway	miR-361-5p	Pum1	Pum2
5HT2 type receptor mediated signaling pathway	3.54×10^{-3}	n.s.	n.s.
5HT4 type receptor mediated signaling pathway	n.s.	n.s.	4.8×10^{-2}
Alpha adrenergic receptor signaling pathway	8.40×10^{-3}	n.s.	n.s.
Alzheimer disease-amyloid secretase pathway	9.73×10^{-3}	2.5×10^{-2}	1.3×10^{-3}
Alzheimer disease-presenilin pathway	n.s.	1.8×10^{-3}	5.4×10^{-6}
Angiogenesis	1.55×10^{-3}	1.0×10^{-7}	1.2×10^{-8}
Apoptosis signaling pathway	4.28×10^{-4}	3.6×10^{-4}	1.4×10^{-6}
Axon guidance mediated by semaphorins	n.s.	2.0×10^{-2}	2.7×10^{-2}
Axon guidance mediated by Slit/Robo	n.s.	5.2×10^{-3}	2.6×10^{-3}
B cell activation	4.85×10^{-5}	3.6×10^{-5}	4.8×10^{-3}
Beta1/2 adrenergic receptor signaling pathway	1.42×10^{-2}	n.s.	3.8×10^{-3}
Beta3 adrenergic receptor signaling pathway	n.s.	n.s.	1.6×10^{-2}
Blood coagulation	n.s.	3.6×10^{-2}	n.s.
Cadherin signaling pathway	n.s.	4.2×10^{-2}	5.7×10^{-3}
Cell cycle	n.s.	3.0×10^{-2}	5.9×10^{-4}
Cholesterol biosynthesis	n.s.	n.s.	4.7×10^{-2}
Circadian clock system	1.10×10^{-2}	3.3×10^{-2}	n.s.
Coenzyme A biosynthesis	n.s.	n.s.	3.2×10^{-2}
Cortocotropin releasing factor receptor signaling pathway	n.s.	n.s.	1.8×10^{-2}
Cytoskeletal regulation by Rho GTPase	n.s.	n.s.	3.8×10^{-2}
De novo purine biosynthesis	n.s.	1.8×10^{-2}	n.s.
DNA replication	n.s.	1.4×10^{-3}	1.1×10^{-4}
EGF receptor signaling pathway	7.20×10^{-6}	5.6×10^{-7}	5.7×10^{-10}
Endothelin signaling pathway	2.62×10^{-4}	n.s.	9.4×10^{-4}
FAS signaling pathway	4.12×10^{-2}	n.s.	1.4×10^{-2}
FGF signaling pathway	2.61×10^{-5}	9.6×10^{-4}	5.6×10^{-9}
Glutamine glutamate conversion	2.14×10^{-2}	n.s.	n.s.
Glycolysis	n.s.	n.s.	5.0×10^{-3}
Hedgehog signaling pathway	n.s.	1.8×10^{-4}	6.0×10^{-5}
Heterotrimeric G-protein signaling pathway-Gq α and Go α mediated pathway	4.03×10^{-2}	n.s.	n.s.
Histamine H1 receptor mediated signaling pathway	1.06×10^{-2}	n.s.	n.s.
Histamine H2 receptor mediated signaling pathway	n.s.	n.s.	2.9×10^{-2}
Huntington disease	n.s.	n.s.	4.0×10^{-6}
Hypoxia response via HIF activation	n.s.	n.s.	2.8×10^{-2}
Inflammation mediated by chemokine and cytokine signaling pathway	4.91×10^{-3}	1.0×10^{-3}	2.1×10^{-2}
Insulin/IGF pathway-MAPKK/MAPK cascade	0.21×10^{-3}	1.9×10^{-3}	3.7×10^{-5}
Insulin/IGF pathway-protein kinase B signaling cascade	n.s.	4.9×10^{-3}	1.6×10^{-4}
Integrin signaling pathway	1.85×10^{-2}	5.7×10^{-5}	1.0×10^{-3}
Interferon-gamma signaling pathway	2.85×10^{-2}	5.4×10^{-4}	6.2×10^{-3}
Interleukin signaling pathway	n.s.	1.6×10^{-5}	6.1×10^{-4}
Metabotropic glutamate receptor group I pathway	1.14×10^{-4}	n.s.	n.s.
Metabotropic glutamate receptor group II pathway	3.96×10^{-2}	n.s.	1.5×10^{-2}

Metabotropic glutamate receptor group III pathway	1.14×10^{-4}	n.s.	2.2×10^{-2}
Muscarinic acetylcholine receptor 1 and 3 signaling pathway	1.95×10^{-2}	n.s.	2.2×10^{-2}
Muscarinic acetylcholine receptor 2 and 4 signaling pathway	n.s.	n.s.	4.4×10^{-2}
Nicotinic acetylcholine receptor signaling pathway	n.s.	n.s.	4.5×10^{-3}
Notch signaling pathway	n.s.	n.s.	1.4×10^{-2}
O-antigen biosynthesis	n.s.	n.s.	4.6×10^{-2}
Oxidative stress response	1.71×10^{-2}	1.0×10^{-3}	5.3×10^{-3}
Oxytocin receptor mediated signaling pathway	8.35×10^{-3}	n.s.	3.4×10^{-2}
p53 pathway	4.32×10^{-5}	1.6×10^{-7}	4.9×10^{-14}
p53 pathway by glucose deprivation	n.s.	n.s.	2.0×10^{-4}
P53 pathway feedback loops 1	2.01×10^{-2}	n.s.	n.s.
p53 pathway feedback loops 2	1.09×10^{-2}	2.2×10^{-2}	1.2×10^{-6}
Parkinson disease	2.42×10^{-2}	4.3×10^{-4}	4.0×10^{-5}
PDGF signaling pathway	7.68×10^{-7}	7.0×10^{-7}	1.6×10^{-12}
PI3 kinase pathway	2.11×10^{-2}	3.7×10^{-4}	2.2×10^{-7}
Ras Pathway	8.18×10^{-4}	1.3×10^{-7}	1.7×10^{-7}
Salvage pyrimidine ribonucleotides	n.s.	2.4×10^{-2}	n.s.
T cell activation	2.19×10^{-6}	8.9×10^{-6}	6.1×10^{-5}
TGF-beta signaling pathway	2.50×10^{-2}	2.6×10^{-4}	9.0×10^{-7}
Thyrotropin-releasing hormone receptor signaling pathway	5.37×10^{-3}	n.s.	1.4×10^{-2}
Toll receptor signaling pathway	n.s.	1.4×10^{-3}	n.s.
Transcription regulation by bZIP transcription factor	n.s.	n.s.	1.1×10^{-2}
Ubiquitin proteasome pathway	n.s.	n.s.	1.2×10^{-6}
Vasopressin synthesis	n.s.	n.s.	1.1×10^{-2}
VEGF signaling pathway	8.46×10^{-3}	1.6×10^{-2}	2.8×10^{-3}
Vitamin B6 metabolism	n.s.	n.s.	2.2×10^{-2}
Wnt signaling pathway	3.18×10^{-3}	4.3×10^{-4}	3.2×10^{-8}

7.7 pTO-HA-Strep-GW-FRT map and sequence

Map of the plasmid pTO-HA-Strep-GW-FRT. The following features are indicated (starting from position 1): A hybrid human cytomegalovirus promoter with tet operator sequences (CMV/tetO promoter) that allows tetracyclin-regulated expression of a gene of interest (tetracyclin displaces the tetracycline repressor protein that binds to the tetO sequences and represses gene expression in the absence of inducer); an HA/StrepIII tandem tag after a Kozak sequence/start codon (not shown); a Gateway cassette consisting of two attachment sites (attR1 and attR2), a chloramphenicol acetyltransferase (CAT) and a 'killer gene' (ccdB) for the convenient recombinase-based cloning of genes of interest using Gateway technology (Invitrogen); a bovine growth hormone (bGH) terminator for the termination of transcription and polyadenylation of transcripts; a Flippase recognition target (FRT) site for genomic integration by Flippase (FLP)-mediated recombination; a hygromycin resistance gene (hygroB) for selection of cells in which recombination occurred (note that the corresponding promoter is present upstream of the FRT site in the target cell lines); an origin of replication (pBR322), an ampicillin resistance marker gene (AmpR) and a corresponding promoter (AmpR promoter) for propagation of the plasmid in *E. coli*. The restriction sites used in this work are indicated (blue). The map was created with PlasMapper (Dong *et al.*, 2004).

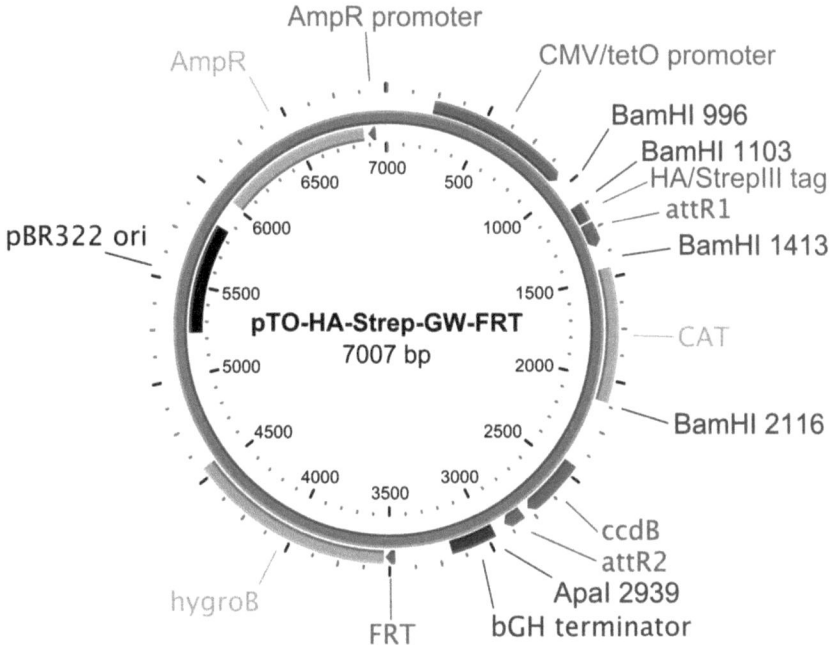

8 Bibliography

Adkins, J. N., Varnum, S. M., Auberry, K. J., Moore, R. J., Angell, N. H., Smith, R. D., Springer, D. L., and Pounds, J. G. (2002). Toward a human blood serum proteome: analysis by multidimensional separation coupled with mass spectrometry. Mol Cell Proteomics 1, 947–955.

Afanasyeva, E. A., Hotz-Wagenblatt, A., Glatting, K.-H., and Westermann, F. (2008). New miRNAs cloned from neuroblastoma. BMC Genomics 9, 52.

Agrawal, D., Hauser, P., McPherson, F., Dong, F., Garcia, A., and Pledger, W. J. (1996). Repression of p27kip1 synthesis by platelet-derived growth factor in BALB/c 3T3 cells. Mol Cell Biol 16, 4327–4336.

Allerson, C. R., Martinez, A., Yikilmaz, E., and Rouault, T. A. (2003). A high-capacity RNA affinity column for the purification of human IRP1 and IRP2 overexpressed in Pichia pastoris. RNA 9, 364–374.

Alon, U. (2007). Network motifs: theory and experimental approaches. Nat Rev Genet 8, 450–461.

Altschul, S. F., Gish, W., Miller, W., Myers, E. W., and Lipman, D. J. (1990). Basic local alignment search tool. J Mol Biol 215, 403–410.

Anantharaman, V., Koonin, E. V., and Aravind, L. (2002). Comparative genomics and evolution of proteins involved in RNA metabolism. Nucleic Acids Res 30, 1427–1464.

Anderson, N. L., and Anderson, N. G. (2002). The human plasma proteome: history, character, and diagnostic prospects. Mol Cell Proteomics 1, 845–867.

Andronescu, M., Condon, A., Hoos, H. H., Mathews, D. H., and Murphy, K. P. (2007). Efficient parameter estimation for RNA secondary structure prediction. Bioinformatics 23, i19–i28.

Andronescu, M., Fejes, A. P., Hutter, F., Hoos, H. H., and Condon, A. (2004). A new algorithm for RNA secondary structure design. J Mol Biol 336, 607–624.

Archer, S. K., Luu, V.-D., de Queiroz, R. A., Brems, S., and Clayton, C. (2009). Trypanosoma brucei PUF9 regulates mRNAs for proteins involved in replicative processes over the cell cycle. PLoS Pathog 5, e1000565.

Bachler, M., Schroeder, R., and von Ahsen, U. (1999). StreptoTag: a novel method for the isolation of RNA-binding proteins. RNA 5, 1509–1516.

Baek, D., Villén, J., Shin, C., Camargo, F. D. F. D., Gygi, S. P. S. P., Bartel, D. P., and Villen, J. (2008). The impact of microRNAs on protein output. Nature 455, 64–71.

Bailey, T. L., Boden, M., Buske, F. A., Frith, M., Grant, C. E., Clementi, L., Ren, J., Li, W. W., and Noble, W. S. (2009). MEME SUITE: tools for motif discovery and searching. Nucleic Acids Res 37, W202–W208.

Bardwell, V. J., and Wickens, M. (1990). Purification of RNA and RNA-protein complexes by an R17 coat protein affinity method. Nucleic Acids Res 18, 6587–6594.

Barrera, L. O., and Ren, B. (2006). The transcriptional regulatory code of eukaryotic cells – insights from genome-wide analysis of chromatin organization and transcription factor binding. Curr Opin Cell Biol 18, 291–298.

Bates, D. O., Cui, T.-G., Doughty, J. M., Winkler, M., Sugiono, M., Shields, J. D., Peat, D., Gillatt, D., and Harper, S. J. (2002). VEGF165b, an inhibitory splice variant of vascular endothelial growth factor, is down-regulated in renal cell carcinoma. Cancer Res 62, 4123–4131.

Beach, D. L., and Keene, J. D. (2008). Ribotrap : targeted purification of RNA-specific RNPs from cell lysates through immunoaffinity precipitation to identify regulatory proteins and RNAs. Methods Mol Biol 419, 69–91.

Bellucci, M., Agostini, F., Masin, M., and Tartaglia, G. G. (2011). Predicting protein associations with long noncoding RNAs. Nat Methods 8, 444–445.

Belotserkovskii, B. P., Liu, R., Tornaletti, S., Krasilnikova, M. M., Mirkin, S. M., and Hanawalt, P. C. (2010). Mechanisms and implications of transcription blockage by guanine-rich DNA sequences. Proc Natl Acad Sci U S A 107, 12816–12821.

Betel, D., Koppal, A., Agius, P., Sander, C., and Leslie, C. (2010). Comprehensive modeling of microRNA targets predicts functional non-conserved and non-canonical sites. Genome Biol 11, R90.

Bhattacharyya, S. N., Habermacher, R., Martine, U., Closs, E. I., and Filipowicz, W. (2006). Relief of microRNA-mediated translational repression in human cells subjected to stress. Cell 125, 1111–1124.

Bindereif, A., and Green, M. R. (1987). An ordered pathway of snRNP binding during mammalian pre-mRNA splicing complex assembly. EMBO J 6, 2415–2424.

Bischofberger, N., Ng, P. G., Webb, T. R., and Matteucci, M. D. (1987). Cleavage of single stranded oligonucleotides by EcoRI restriction endonuclease. Nucleic Acids Res 15, 709–716.

Blencowe, B. J., and Barabino, S. M. (1995). Antisense affinity depletion of RNP particles. Application to spliceosomal snRNPs. Methods Mol Biol 37, 67–76.

Bourdeau, V., Ferbeyre, G., Pageau, M., Paquin, B., and Cedergren, R. (1999). The distribution of RNA motifs in natural sequences. Nucleic Acids Res 27, 4457–4467.

Bowden, J., Brennan, P. A., Umar, T., and Cronin, A. (2002). Expression of vascular endothelial growth factor in basal cell carcinoma and cutaneous squamous cell carcinoma of the head and neck. J Cutan Pathol 29, 585–589.

Brenner, S., Jacob, F., and Meselson, M. (1961). An unstable intermediate carrying information from genes to ribosomes for protein synthesis. Nature 190, 576–581.

Brickner, A. G., Warren, E. H., Caldwell, J. A., Akatsuka, Y., Golovina, T. N., Zarling, A. L., Shabanowitz, J., Eisenlohr, L. C., Hunt, D. F., Engelhard, V. H., et al. (2001). The immunogenicity of a new human minor histocompatibility antigen results from differential antigen processing. J Exp Med 193, 195–206.

Brodsky, A. S., and Silver, P. A. (2000). Pre-mRNA processing factors are required for nuclear export. RNA 6, 1737–1749.

Butter, F., Scheibe, M., Mörl, M., and Mann, M. (2009). Unbiased RNA-protein interaction screen by quantitative proteomics. Proc Natl Acad Sci U S A 106, 10626–10631.

Böck-Taferner, P., and Wank, H. (2004). GAPDH enhances group II intron splicing in vitro. Biol Chem 385, 615–621.

Calin, G. A., and Croce, C. M. (2006). MicroRNA signatures in human cancers. Nat Rev Cancer 6, 857–866.

Caputi, M., Mayeda, A., Krainer, A. R., and Zahler, A. M. (1999). hnRNP A/B proteins are required for inhibition of HIV-1 pre-mRNA splicing. EMBO J 18, 4060–4067.

Carmeliet, P. (2005). Angiogenesis in life, disease and medicine. Nature 438, 932–936.

Chang, H.-Y., Fan, C.-C., Chu, P.-C., Hong, B.-E., Lee, H. J., and Chang, M.-S. (2011). hPuf-A/KIAA0020 modulates PARP-1 cleavage upon genotoxic stress. Cancer Res 71, 1126–1134.

Chang, S.-H., and Hla, T. (2011). Gene regulation by RNA binding proteins and microRNAs in angiogenesis. Trends Mol Med 17, 650–658.

Chen, S. J., and Dill, K. A. (2000). RNA folding energy landscapes. Proc Natl Acad Sci U S A 97, 646–651.

Chendrimada, T. P., Gregory, R. I., Kumaraswamy, E., Norman, J., Cooch, N., Nishikura, K., and Shiekhattar, R. (2005). TRBP recruits the Dicer complex to Ago2 for microRNA processing and gene silencing. Nature 436, 740–744.

Cheong, C.-G., and Hall, T. M. T. (2006). Engineering RNA sequence specificity of Pumilio repeats. Proc Natl Acad Sci U S A 103, 13635–13639.

Chi, S. W., Zang, J. B., Mele, A., and Darnell, R. B. (2009). Argonaute HITS-CLIP decodes microRNA-mRNA interaction maps. Nature 460, 479–486.

Cho, S., Kim, J. H., Back, S. H., and Jang, S. K. (2005). Polypyrimidine tract-binding protein enhances the internal ribosomal entry site-dependent translation of p27Kip1 mRNA and modulates transition from G1 to S phase. Mol Cell Biol 25, 1283–1297.

Chu, S., DeRisi, J., Eisen, M., Mulholland, J., Botstein, D., Brown, P. O., and Herskowitz, I. (1998). The transcriptional program of sporulation in budding yeast. Science 282, 699–705.

Ciais, D., Cherradi, N., Bailly, S., Grenier, E., Berra, E., Pouyssegur, J., Lamarre, J., and

Feige, J.-J. (2004). Destabilization of vascular endothelial growth factor mRNA by the zinc-finger protein TIS11b. Oncogene 23, 8673–8680.

Claffey, K. P., Shih, S. C., Mullen, A., Dziennis, S., Cusick, J. L., Abrams, K. R., Lee, S. W., and Detmar, M. (1998). Identification of a human VPF/VEGF 3' untranslated region mediating hypoxia-induced mRNA stability. Mol Biol Cell 9, 469–481.

Coles, L. S., Bartley, M. A., Bert, A., Hunter, J., Polyak, S., Diamond, P., Vadas, M. A., and Goodall, G. J. (2004). A multi-protein complex containing cold shock domain (Y-box) and polypyrimidine tract binding proteins forms on the vascular endothelial growth factor mRNA. Potential role in mRNA stabilization. Eur Journal Biochem 271, 648–660.

Corcoran, D. L., Georgiev, S., Mukherjee, N., Gottwein, E., Skalsky, R. L., Keene, J. D., and Ohler, U. (2011). PARalyzer: definition of RNA binding sites from PAR-CLIP short-read sequence data. Genome Biol 12, R79.

Crick, F. H., Barnett, L., Brenner, S., and Watts-Tobin, R. J. (1961). General nature of the genetic code for proteins. Nature 192, 1227–1232.

Cuesta, R., Martínez-Sánchez, A., and Gebauer, F. (2009). miR-181a regulates cap-dependent translation of p27(kip1) mRNA in myeloid cells. Mol Cell Biol 29, 2841–2851.

Czaplinski, K., Köcher, T., Schelder, M., Segref, A., Wilm, M., and Mattaj, I. W. (2005). Identification of 40LoVe, a Xenopus hnRNP D family protein involved in localizing a TGF-beta-related mRNA during oogenesis. Dev Cell 8, 505–515.

Dangerfield, J. A., Windbichler, N., Salmons, B., Günzburg, W. H., and Schröder, R. (2006). Enhancement of the StreptoTag method for isolation of endogenously expressed proteins with complex RNA binding targets. Electrophoresis 27, 1874–1877.

Darnell, J. C., Van Driesche, S. J., Zhang, C., Hung, K. Y. S., Mele, A., Fraser, C. E., Stone, E. F., Chen, C., Fak, J. J., Chi, S. W., et al. (2011). FMRP Stalls Ribosomal Translocation on mRNAs Linked to Synaptic Function and Autism. Cell 146, 247–261.

Darnell, R. B. (2010). HITS-CLIP: panoramic views of protein-RNA regulation in living cells. Wiley Interdiscip Rev RNA 1, 266–286.

Das, R., Zhou, Z., and Reed, R. (2000). Functional association of U2 snRNP with the ATP-independent spliceosomal complex E. Mol Cell 5, 779–787.

Deckert, J., Hartmuth, K., Boehringer, D., Behzadnia, N., Will, C. L., Kastner, B., Stark, H., Urlaub, H., and Lührmann, R. (2006). Protein composition and electron microscopy structure of affinity-purified human spliceosomal B complexes isolated under physiological conditions. Mol Cell Biol 26, 5528–5543.

Denli, A. M., Tops, B. B. J., Plasterk, R. H. a, Ketting, R. F., and Hannon, G. J. (2004). Processing of primary microRNAs by the Microprocessor complex. Nature 432, 231–235.

Detmar, M. (2000). The role of VEGF and thrombospondins in skin angiogenesis. J Dermatol Sci 24 Suppl 1, S78–S84.

Detmar, M., Brown, L. F., Berse, B., Jackman, R. W., Elicker, B. M., Dvorak, H. F., and Claffey, K. P. (1997). Hypoxia regulates the expression of vascular permeability factor/vascular endothelial growth factor (VPF/VEGF) and its receptors in human skin. J Invest Dermatol 108, 263–268.

Detmar, M., Brown, L. F., Claffey, K. P., Yeo, K. T., Kocher, O., Jackman, R. W., Berse, B., and Dvorak, H. F. (1994). Overexpression of vascular permeability factor/vascular endothelial growth factor and its receptors in psoriasis. J Exp Med 180, 1141–1146.

Dibbens, J. A., Miller, D. L., Damert, A., Risau, W., Vadas, M. A., and Goodall, G. J. (1999). Hypoxic regulation of vascular endothelial growth factor mRNA stability requires the cooperation of multiple RNA elements. Mol Biol Cell 10, 907–919.

Dibbens, J. A., Polyak, S. W., Damert, A., Risau, W., Vadas, M. a, and Goodall, G. J. (2001). Nucleotide sequence of the mouse VEGF 3'UTR and quantitative analysis of sites of polyadenylation. Biochim Biophy Acta 1518, 57–62.

Djuranovic, S., Nahvi, A., and Green, R. (2011). A parsimonious model for gene regulation by miRNAs. Science 331, 550–553.

Dong, S., Wang, Y., Cassidy-Amstutz, C., Lu, G., Bigler, R., Jezyk, M. R., Li, C., Hall, T. M. T., and Wang, Z. (2011). Specific and modular binding code for cytosine recognition in Pumilio/FBF (PUF) RNA-binding domains. J Biol Chem 286, 26732–26742.

Dong, X., Stothard, P., Forsythe, I. J., and Wishart, D. S. (2004). PlasMapper: a web server for drawing and auto-annotating plasmid maps. Nucleic Acids Res 32, W660–W664.

Dziunycz, P., Iotzova-Weiss, G., Eloranta, J. J., Läuchli, S., Hafner, J., French, L. E., and Hofbauer, G. F. L. (2010). Squamous cell carcinoma of the skin shows a distinct microRNA profile modulated by UV radiation. J Invest Dermatol 130, 2686–2689.

Ellington, A. D., and Szostak, J. W. (1990). In vitro selection of RNA molecules that bind specific ligands. Nature 346, 818–822.

Erickson, S. L., and Lykke-Andersen, J. (2011). Cytoplasmic mRNP granules at a glance. J Cell Sci 124, 293–297.

Farazi, T. a, Spitzer, J. I., Morozov, P., and Tuschl, T. (2011). miRNAs in human cancer. J Pathol 223, 102–115.

Felicetti, F., Errico, M. C., Bottero, L., Segnalini, P., Stoppacciaro, A., Biffoni, M., Felli, N., Mattia, G., Petrini, M., Colombo, M. P., et al. (2008). The promyelocytic leukemia zinc finger-microRNA-221/-222 pathway controls melanoma progression through multiple oncogenic mechanisms. Cancer Res 68, 2745–2754.

Ferrara, N., Gerber, H.-P., and LeCouter, J. (2003). The biology of VEGF and its receptors. Nat Med 9, 669–676.

Ferrara, N., and Henzel, W. J. (1989). Pituitary follicular cells secrete a novel heparin-binding growth factor specific for vascular endothelial cells. Biochem Biophys Res Commun 161, 851–858.

Filipovska, A., Razif, M. F. M., Nygård, K. K. a, and Rackham, O. (2011). A universal code for RNA recognition by PUF proteins. Nat Chem Biol 7, 425–427.

Filipovska, A., and Rackham, O. (2011). Designer RNA-binding proteins: New tools for manipulating the transcriptome. RNA Biol 8.

Filipowicz, W., Bhattacharyya, S. N., and Sonenberg, N. (2008). Mechanisms of post-transcriptional regulation by microRNAs: are the answers in sight? Nat Rev Genet 9, 102–114.

Folkman, J. (1990). What is the evidence that tumors are angiogenesis dependent? J Natl Cancer Inst 82, 4–6.

Folkman, J., Merler, E., Abernathy, C., and Williams, G. (1971). Isolation of a tumor factor responsible for angiogenesis. J Exp Med 133, 275–288.

Fornari, F., Gramantieri, L., Ferracin, M., Veronese, A., Sabbioni, S., Calin, G. A., Grazi, G. L., Giovannini, C., Croce, C. M., Bolondi, L., et al. (2008). MiR-221 controls CDKN1C/p57 and CDKN1B/p27 expression in human hepatocellular carcinoma. Oncogene 27, 5651–5661.

Frank, S., Hübner, G., Breier, G., Longaker, M. T., Greenhalgh, D. G., and Werner, S. (1995). Regulation of vascular endothelial growth factor expression in cultured keratinocytes. Implications for normal and impaired wound healing. J Biol Chem 270, 12607–12613.

Friedman, R. C., Farh, K. K.-how, Burge, C. B., and Bartel, D. P. (2009). Most mammalian mRNAs are conserved targets of microRNAs. Genome Res 19, 92–105.

Fujita, P. A., Rhead, B., Zweig, A. S., Hinrichs, A. S., Karolchik, D., Cline, M. S., Goldman, M., Barber, G. P., Clawson, H., Coelho, A., et al. (2011). The UCSC Genome Browser database: update 2011. Nucleic Acids Res 39, D876–D882.

Gaidatzis, D., van Nimwegen, E., Hausser, J., Zavolan, M., Nimwegen, V., E., H., and J (2007). Inference of miRNA targets using evolutionary conservation and pathway analysis. BMC Bioinformatics 8, 69.

Galardi, S., Mercatelli, N., Giorda, E., Massalini, S., Frajese, G. V., Ciafrè, S. A., and Farace, M. G. (2007). miR-221 and miR-222 expression affects the proliferation potential of human prostate carcinoma cell lines by targeting p27Kip1. J Biol Chem 282, 23716–23724.

Galgano, A. (2010). Pum1 affects VEGF-A expression. In "Comparative analysis of mRNA targets for human PUF-family-RNA binding proteins" (ETH Zurich, Diss. No. 18803), pp. 139–169.

Galgano, A., Forrer, M., Jaskiewicz, L., Kanitz, A., Zavolan, M., and Gerber, A. P. (2008). Comparative analysis of mRNA targets for human PUF-family proteins suggests extensive interaction with the miRNA regulatory system. PLoS One 3, e3164.

Galperin, M. Y., and Cochrane, G. R. (2009). Nucleic Acids Research annual Database Issue and the NAR online Molecular Biology Database Collection in 2009. Nucleic Acids Res

37, D1–D4.

Gerber, A. P., Herschlag, D., and Brown, P. O. (2004a). Extensive association of functionally and cytotopically related mRNAs with Puf family RNA-binding proteins in yeast. PLoS Biol 2, E79.

Gerber, A. P., Luschnig, S., Krasnow, M. A., Brown, P. O., and Herschlag, D. (2006). Genome-wide identification of mRNAs associated with the translational regulator PUMILIO in Drosophila melanogaster. Proc Natl Acad Sci U S A 103, 4487–4492.

Gerber, C. A., Relich, A., and Driscoll, D. M. (2004b). Isolation of an mRNA-binding protein involved in C-to-U editing. Methods Mol Biol 265, 239–249.

Godfried Sie, C. P., and Kuchka, M. (2011). RNA editing adds flavor to complexity. Biochemistry (Mosc) 76, 869–881.

Goldberg-Cohen, I., Furneauxb, H., and Levy, A. P. (2002). A 40-bp RNA element that mediates stabilization of vascular endothelial growth factor mRNA by HuR. J Biol Chem 277, 13635–13640.

Gonsalvez, G. B., Little, J. L., and Long, R. M. (2004). ASH1 mRNA anchoring requires reorganization of the Myo4p-She3p-She2p transport complex. J Biol Chem 279, 46286–46294.

Grabowski, P. J., and Sharp, P. A. (1986). Affinity chromatography of splicing complexes: U2, U5, and U4 + U6 small nuclear ribonucleoprotein particles in the spliceosome. Science 233, 1294–1299.

Graham, F. L., Smiley, J., Russell, W. C., and Nairn, R. (1977). Characteristics of a human cell line transformed by DNA from human adenovirus type 5. J Gen Virol 36, 59–74.

Green, N. M. (1990). Avidin and streptavidin. Methods Enzymol 184, 51–67.

Greenberg, J. R. (1979). Ultraviolet light-induced crosslinking of mRNA to proteins. Nucleic Acids Res 6, 715–732.

Gregory, R. I., Chendrimada, T. P., Cooch, N., and Shiekhattar, R. (2005). Human RISC couples microRNA biogenesis and posttranscriptional gene silencing. Cell 123, 631–640.

Gregory, R. I., Yan, K.-P., Amuthan, G., Chendrimada, T., Doratotaj, B., Cooch, N., and Shiekhattar, R. (2004). The Microprocessor complex mediates the genesis of microRNAs. Nature 432, 235–240.

Griffiths-Jones, S., Saini, H. K., van Dongen, S., and Enright, A. J. (2008). miRBase: tools for microRNA genomics. Nucleic Acids Res 36, D154–D158.

Gros, F., Hiatt, H., Gilbert, W., Kurland, C. G., Risebrough, R. W., and Watson, J. D. (1961). Unstable ribonucleic acid revealed by pulse labelling of Escherichia coli. Nature 190, 581–585.

Gruber, A. R., Lorenz, R., Bernhart, S. H., Neuböck, R., and Hofacker, I. L. (2008). The

Vienna RNA websuite. Nucleic Acids Res 36, W70–W74.

Gstaiger, M., and Aebersold, R. (2009). Applying mass spectrometry-based proteomics to genetics, genomics and network biology. Nat Rev Genet 10, 617–627.

Guo, S.-L., Peng, Z., Yang, X., Fan, K.-J., Ye, H., Li, Z.-H., Wang, Y., Xu, X.-L., Li, J., Wang, Y.-L., et al. (2011). miR-148a promoted cell proliferation by targeting p27 in gastric cancer cells. Int J Biol Sci 7, 567–574.

Gygi, S. P., Rochon, Y., Franza, B. R., and Aebersold, R. (1999). Correlation between protein and mRNA abundance in yeast. Mol Cell Biol 19, 1720–1730.

Hafner, M., Landthaler, M., Burger, L., Khorshid, M., Hausser, J., Berninger, P., Rothballer, A., Ascano, M., Jungkamp, A.-C., Munschauer, M., et al. (2010a). PAR-CliP – a method to identify transcriptome-wide the binding sites of RNA binding proteins. J Vis Exp.

Hafner, M., Landthaler, M., Burger, L., Khorshid, M., Hausser, J., Berninger, P., Rothballer, A., Ascano, M., Jungkamp, A.-C., Munschauer, M., et al. (2010b). Transcriptome-wide identification of RNA-binding protein and microRNA target sites by PAR-CLIP. Cell 141, 129–141.

Halbeisen, R. E., Galgano, A., Scherrer, T., and Gerber, A. P. (2008). Post-transcriptional gene regulation: from genome-wide studies to principles. Cell Mol Life Sci 65, 798–813.

Hamasaki, K., Killian, J., Cho, J., and Rando, R. R. (1998). Minimal RNA constructs that specifically bind aminoglycoside antibiotics with high affinities. Biochemistry 37, 656–663.

Harbison, C. T., Gordon, D. B., Lee, T. I., Rinaldi, N. J., Macisaac, K. D., Danford, T. W., Hannett, N. M., Tagne, J.-B., Reynolds, D. B., Yoo, J., et al. (2004). Transcriptional regulatory code of a eukaryotic genome. Nature 431, 99–104.

Harper, S. J., and Bates, D. O. (2008). VEGF-A splicing: the key to anti-angiogenic therapeutics? Nat Rev Cancer 8, 880–887.

Hartmuth, K., Urlaub, H., Vornlocher, H.-P., Will, C. L., Gentzel, M., Wilm, M., and Lührmann, R. (2002). Protein composition of human prespliceosomes isolated by a tobramycin affinity-selection method. Proc Natl Acad Sci U S A 99, 16719–16724.

Hartmuth, K., Vornlocher, H.-P., and Lührmann, R. (2004). Tobramycin affinity tag purification of spliceosomes. Methods Mol Biol 257, 47–64.

Hendrickson, D. G., Hogan, D. J., Herschlag, D., Ferrell, J. E., and Brown, P. O. (2008). Systematic identification of mRNAs recruited to argonaute 2 by specific microRNAs and corresponding changes in transcript abundance. PLoS One 3, e2126.

Hengst, L., and Reed, S. I. (1996). Translational control of p27Kip1 accumulation during the cell cycle. Science 271, 1861–1864.

Hershko, D. D. (2010). Cyclin-dependent kinase inhibitor p27 as a prognostic biomarker and potential cancer therapeutic target. Future Oncol 6, 1837–1847.

Hieronymus, H., and Silver, P. A. (2004). A systems view of mRNP biology. Genes Dev 18, 2845–2860.

Hoeben, A., Landuyt, B., Highley, M. S., Wildiers, H., Van Oosterom, A. T., and De Bruijn, E. A. (2004). Vascular endothelial growth factor and angiogenesis. Pharmacol Rev 56, 549–580.

Hofacker, I. L., Flamm, C., Heine, C., Wolfinger, M. T., Scheuermann, G., and Stadler, P. F. (2010). BarMap: RNA folding on dynamic energy landscapes. RNA 16, 1308–1316.

Hofbauer, G. F. L., Bouwes Bavinck, J. N., and Euvrard, S. (2010). Organ transplantation and skin cancer: basic problems and new perspectives. Exp Dermatol 19, 473–482.

Hogan, D. J., Riordan, D. P., Gerber, A. P., Herschlag, D., and Brown, P. O. (2008). Diverse RNA-binding proteins interact with functionally related sets of RNAs, suggesting an extensive regulatory system. PLoS Biol 6, e255.

Hogg, J. R., and Collins, K. (2007a). Human Y5 RNA specializes a Ro ribonucleoprotein for 5S ribosomal RNA quality control. Genes Dev 21, 3067–3072.

Hogg, J. R., and Collins, K. (2007b). RNA-based affinity purification reveals 7SK RNPs with distinct composition and regulation. RNA 13, 868–880.

Houck, K. A., Ferrara, N., Winer, J., Cachianes, G., Li, B., and Leung, D. W. (1991). The vascular endothelial growth factor family: identification of a fourth molecular species and characterization of alternative splicing of RNA. Mol Endocrinol 5, 1806–1814.

Hua, Z., Lv, Q., Ye, W., Wong, C.-K. A., Cai, G., Gu, D., Ji, Y., Zhao, C., Wang, J., Yang, B. B., et al. (2006). MiRNA-directed regulation of VEGF and other angiogenic factors under hypoxia. PLoS One 1, e116.

Huang, D. W., Sherman, B. T., Stephens, R., Baseler, M. W., Lane, H. C., and Lempicki, R. A. (2008). DAVID gene ID conversion tool. Bioinformation 2, 428–430.

Huez, I., Bornes, S., Bresson, D., Créancier, L., and Prats, H. (2001). New vascular endothelial growth factor isoform generated by internal ribosome entry site-driven CUG translation initiation. Mol Endocrinol 15, 2197–2210.

Höck, J., and Meister, G. (2008). The Argonaute protein family. Genome Biol 9, 210.

Iioka, H., Loiselle, D., Haystead, T. A., and Macara, I. G. (2011). Efficient detection of RNA-protein interactions using tethered RNAs. Nucleic Acids Res 39, e53.

Iyer, V. R., Horak, C. E., Scafe, C. S., Botstein, D., Snyder, M., and Brown, P. O. (2001). Genomic binding sites of the yeast cell-cycle transcription factors SBF and MBF. Nature 409, 533–538.

Jacob, F., and Monod, J. (1961). Genetic regulatory mechanisms in the synthesis of proteins. J Mol Biol 3, 318–356.

Jafarifar, F., Yao, P., Eswarappa, S. M., and Fox, P. L. (2011). Repression of VEGFA by CA-rich element-binding microRNAs is modulated by hnRNP L. EMBO J 30, 1324–1334.

Jiang, L., Suri, A. K., Fiala, R., and Patel, D. J. (1997). Saccharide-RNA recognition in an aminoglycoside antibiotic-RNA aptamer complex. Chem Biol 4, 35–50.

Jiang, L., and Patel, D. J. (1998). Solution structure of the tobramycin-RNA aptamer complex. Nat Struct Biol 5, 769–774.

Jurica, M. S., Licklider, L. J., Gygi, S. R., Grigorieff, N., and Moore, M. J. (2002). Purification and characterization of native spliceosomes suitable for three-dimensional structural analysis. RNA 8, 426–439.

Jurica, M. S., and Moore, M. J. (2002). Capturing splicing complexes to study structure and mechanism. Methods 28, 336–345.

Kaminski, A., Ostareck, D. H., Standart, N. M., and Jackson, R. J. (1998). Affinity methods for isolating RNA binding proteins. In RNA:Protein Interactions: A Practical Approach, C. W. J. Smith, ed. (New York: Oxford Univsersity Press), pp. 137–160.

Kanehisa, M., Goto, S., Furumichi, M., Tanabe, M., and Hirakawa, M. (2010). KEGG for representation and analysis of molecular networks involving diseases and drugs. Nucleic Acids Res 38, D355–D360.

Kanitz, A., and Gerber, A. P. (2010). Circuitry of mRNA regulation. Wiley Interdiscip Rev Syst Biol Med 2, 245–251.

Karaa, Z. S., Iacovoni, J. S., Bastide, A., Lacazette, E., Touriol, C., and Prats, H. (2009). The VEGF IRESes are differentially susceptible to translation inhibition by miR-16. RNA 15, 249–254.

Karginov, F. V., Conaco, C., Xuan, Z., Schmidt, B. H., Parker, J. S., Mandel, G., and Hannon, G. J. (2007). A biochemical approach to identifying microRNA targets. Proc Natl Acad Sci U S A 104, 19291–19296.

Keck, P. J., Hauser, S. D., Krivi, G., Sanzo, K., Warren, T., Feder, J., and Connolly, D. T. (1989). Vascular permeability factor, an endothelial cell mitogen related to PDGF. Science 246, 1309–1312.

Kedde, M., van Kouwenhove, M., Zwart, W., Oude Vrielink, J. A. F., Elkon, R., and Agami, R. (2010). A Pumilio-induced RNA structure switch in p27-3' UTR controls miR-221 and miR-222 accessibility. Nat Cell Biol 12, 1014–1020.

Kedde, M., Strasser, M. J., Boldajipour, B., Oude Vrielink, J. A. F., Slanchev, K., le Sage, C., Nagel, R., Voorhoeve, P. M., van Duijse, J., Ørom, U. A., et al. (2007). RNA-binding protein Dnd1 inhibits microRNA access to target mRNA. Cell 131, 1273–1286.

Kedersha, N., and Anderson, P. (2007). Mammalian stress granules and processing bodies. Methods Enzymol 431, 61–81.

Keefe, A. D., Wilson, D. S., Seelig, B., and Szostak, J. W. (2001). One-step purification of recombinant proteins using a nanomolar-affinity streptavidin-binding peptide, the SBP-

Tag. Protein Expr Purif 23, 440–446.

Keene, J. D. (2007). RNA regulons: coordination of post-transcriptional events. Nat Rev Genet 8, 533–543.

Keene, J. D. (2001). Ribonucleoprotein infrastructure regulating the flow of genetic information between the genome and the proteome. Proc Natl Acad Sci U S A 98, 7018–7024.

Kershner, A. M., and Kimble, J. (2010). Genome-wide analysis of mRNA targets for Caenorhabditis elegans FBF, a conserved stem cell regulator. Proc Natl Acad Sci U S A 107, 3936–3941.

Khorshid, M., Rodak, C., and Zavolan, M. (2011). CLIPZ: a database and analysis environment for experimentally determined binding sites of RNA-binding proteins. Nucleic Acids Res 39, D245–D252.

Kim, N., and Jinks-Robertson, S. (2011). Guanine repeat-containing sequences confer transcription-dependent instability in an orientation-specific manner in yeast. DNA Repair (Amst) 10, 953–960.

Kishore, S., Jaskiewicz, L., Burger, L., Hausser, J., Khorshid, M., and Zavolan, M. (2011). A quantitative analysis of CLIP methods for identifying binding sites of RNA-binding proteins. Nat Methods.

van Kouwenhove, M., Kedde, M., and Agami, R. (2011). MicroRNA regulation by RNA-binding proteins and its implications for cancer. Nat Rev Cancer 11, 644–656.

Kozomara, A., and Griffiths-Jones, S. (2011). miRBase: integrating microRNA annotation and deep-sequencing data. Nucleic Acids Res 39, D152–D157.

Krol, J., Loedige, I., and Filipowicz, W. (2010). The widespread regulation of microRNA biogenesis, function and decay. Nat Rev Genet 11, 597–610.

Kula, A., Guerra, J., Knezevich, A., Kleva, D., Myers, M. P., and Marcello, A. (2011). Characterization of the HIV-1 RNA associated proteome identifies Matrin 3 as a nuclear cofactor of Rev function. Retrovirology 8, 60.

Kullmann, M., Göpfert, U., Siewe, B., and Hengst, L. (2002). ELAV/Hu proteins inhibit p27 translation via an IRES element in the p27 5'UTR. Genes Dev 16, 3087–3099.

Kurosu, T., Ohga, N., Hida, Y., Maishi, N., Akiyama, K., Kakuguchi, W., Kuroshima, T., Kondo, M., Akino, T., Totsuka, Y., et al. (2011). HuR keeps an angiogenic switch on by stabilising mRNA of VEGF and COX-2 in tumour endothelium. Br J Cancer 104, 819–829.

König, J., Zarnack, K., Rot, G., Curk, T., Kayikci, M., Zupan, B., Turner, D. J., Luscombe, N. M., and Ule, J. (2010). iCLIP reveals the function of hnRNP particles in splicing at individual nucleotide resolution. Nat Struct Mol Biol 17, 909–915.

König, J., Zarnack, K., Rot, G., Curk, T., Kayikci, M., Zupan, B., Turner, D. J., Luscombe, N.

M., and Ule, J. (2011). iCLIP – transcriptome-wide mapping of protein-RNA interactions with individual nucleotide resolution. J Vis Exp.

Lagos-Quintana, M., Rauhut, R., Lendeckel, W., and Tuschl, T. (2001). Identification of novel genes coding for small expressed RNAs. Science 294, 853–858.

Lamond, a I., Sproat, B., Ryder, U., and Hamm, J. (1989). Probing the structure and function of U2 snRNP with antisense oligonucleotides made of 2'-OMe RNA. Cell 58, 383–390.

Lamond, A. I., and Sproat, B. S. (1994). Isolation and characterisation of ribonucleoprotein complexes. In RNA Processing: A Practical Approach, Volume 1, D. Hames and S. Higgins, eds. (New York: Oxford Univsersity Press), pp. 103–140.

Landthaler, M., Gaidatzis, D., Rothballer, A., Chen, P. Y., Soll, S. J., Dinic, L., Ojo, T., Hafner, M., Zavolan, M., and Tuschl, T. (2008). Molecular characterization of human Argonaute-containing ribonucleoprotein complexes and their bound target mRNAs. RNA 14, 2580–2596.

Lange, T., Guttmann-Raviv, N., Baruch, L., Machluf, M., and Neufeld, G. (2003). VEGF162, a new heparin-binding vascular endothelial growth factor splice form that is expressed in transformed human cells. J Biol Chem 278, 17164–17169.

Langland, J. O., Pettiford, S. M., and Jacobs, B. L. (1995). Nucleic acid affinity chromatography: preparation and characterization of double-stranded RNA agarose. Protein Expr Purif 6, 25–32.

Larocque, D., Galarneau, A., Liu, H.-N., Scott, M., Almazan, G., and Richard, S. (2005). Protection of p27(Kip1) mRNA by quaking RNA binding proteins promotes oligodendrocyte differentiation. Nat Neurosci 8, 27–33.

Lasko, P. (2003). Gene regulation at the RNA layer: RNA binding proteins in intercellular signaling networks. Sci STKE 2003, RE6.

Lau, N. C., Lim, L. P., Weinstein, E. G., and Bartel, D. P. (2001). An abundant class of tiny RNAs with probable regulatory roles in Caenorhabditis elegans. Science 294, 858–862.

Lebedeva, S., Jens, M., Theil, K., Schwanhäusser, B., Selbach, M., Landthaler, M., and Rajewsky, N. (2011). Transcriptome-wide Analysis of Regulatory Interactions of the RNA-Binding Protein HuR. Mol Cell 43, 1–13.

Lee, J. F., Hesselberth, J. R., Meyers, L. A., and Ellington, A. D. (2004). Aptamer database. Nucleic Acids Res 32, D95–D100.

Lee, R. C., and Ambros, V. (2001). An extensive class of small RNAs in Caenorhabditis elegans. Science 294, 862–864.

Lee, R. C., Feinbaum, R. L., and Ambros, V. (1993). The C. elegans heterochronic gene lin-4 encodes small RNAs with antisense complementarity to lin-14. Cell 75, 843–854.

Lee, T. I., Rinaldi, N. J., Robert, F., Odom, D. T., Bar-Joseph, Z., Gerber, G. K., Hannett, N. M., Harbison, C. T., Thompson, C. M., Simon, I., et al. (2002). Transcriptional regulatory networks in Saccharomyces cerevisiae. Science 298, 799–804.

Lei, J., Jiang, A., and Pei, D. (1998). Identification and characterization of a new splicing variant of vascular endothelial growth factor: VEGF183. Biochim Biophy Acta 1443, 400–406.

Lei, Z., Li, B., Yang, Z., Fang, H., Zhang, G.-M., Feng, Z.-H., and Huang, B. (2009). Regulation of HIF-1alpha and VEGF by miR-20b tunes tumor cells to adapt to the alteration of oxygen concentration. PLoS One 4, e7629.

Lemay, V., Hossain, A., Osheim, Y. N., Beyer, A. L., and Dragon, F. (2011). Identification of novel proteins associated with yeast snR30 small nucleolar RNA. Nucleic Acids Res 3000, 1–12.

Leulliot, N., and Varani, G. (2001). Current topics in RNA-protein recognition: control of specificity and biological function through induced fit and conformational capture. Biochemistry 40, 7947–7956.

Leung, D. W., Cachianes, G., Kuang, W. J., Goeddel, D. V., and Ferrara, N. (1989). Vascular endothelial growth factor is a secreted angiogenic mitogen. Science 246, 1306–1309.

Levy, N. S., Chung, S., Furneaux, H., and Levy, A. P. (1998). Hypoxic stabilization of vascular endothelial growth factor mRNA by the RNA-binding protein HuR. J Biol Chem 273, 6417–6423.

Levy, N. S., Goldberg, M. A., and Levy, A. P. (1997). Sequencing of the human vascular endothelial growth factor (VEGF) 3' untranslated region (UTR): conservation of five hypoxia-inducible RNA-protein binding sites. Biochim Biophy Acta 1352, 167–173.

Lewis, B. P., Burge, C. B., and Bartel, D. P. (2005). Conserved seed pairing, often flanked by adenosines, indicates that thousands of human genes are microRNA targets. Cell 120, 15–20.

Li, Y., and Altman, S. (2002). Partial reconstitution of human RNase P in HeLa cells between its RNA subunit with an affinity tag and the intact protein components. Nucleic Acids Res 30, 3706–3711.

Li, Y.-L., Ye, F., Hu, Y., Lu, W.-G., and Xie, X. (2009). Identification of suitable reference genes for gene expression studies of human serous ovarian cancer by real-time polymerase chain reaction. Anal Biochem 394, 110–116.

Liang, Y., Li, X.-Y., Rebar, E. J., Li, P., Zhou, Y., Chen, B., Wolffe, A. P., and Case, C. C. (2002). Activation of vascular endothelial growth factor A transcription in tumorigenic glioblastoma cell lines by an enhancer with cell type-specific DNase I accessibility. J Biol Chem 277, 20087–20094.

Licatalosi, D. D., Mele, A., Fak, J. J., Ule, J., Kayikci, M., Chi, S. W., Clark, T. a, Schweitzer, A. C., Blume, J. E., Wang, X., et al. (2008). HITS-CLIP yields genome-wide insights into brain alternative RNA processing. Nature 456, 464–469.

Lieb, J. D., Liu, X., Botstein, D., and Brown, P. O. (2001). Promoter-specific binding of Rap1 revealed by genome-wide maps of protein-DNA association. Nat Genet 28, 327–334.

Lingner, J., and Cech, T. R. (1996). Purification of telomerase from Euplotes aediculatus: requirement of a primer 3' overhang. Proc Natl Acad Sci U S A 93, 10712–10717.

Liu, B., Peng, X.-C., Zheng, X.-L., Wang, J., and Qin, Y.-W. (2009). MiR-126 restoration down-regulate VEGF and inhibit the growth of lung cancer cell lines in vitro and in vivo. Lung Cancer 66, 169–175.

Liu, Z., Dong, Z., Han, B., Yang, Y., Liu, Y., and Zhang, J.-T. (2005). Regulation of expression by promoters versus internal ribosome entry site in the 5'-untranslated sequence of the human cyclin-dependent kinase inhibitor p27kip1. Nucleic Acids Res 33, 3763–3771.

Locker, N., Easton, L. E., and Lukavsky, P. J. (2006). Affinity purification of eukaryotic 48S initiation complexes. RNA 12, 683–690.

Locker, N., and Lukavsky, P. J. (2007). A practical approach to isolate 48S complexes: affinity purification and analyses. Methods Enzymol 429, 83–104.

Lohmann, C. M., and Solomon, A. R. (2001). Clinicopathologic variants of cutaneous squamous cell carcinoma. Adv Anat Pathol 8, 27–36.

Long, J., Wang, Y., Wang, W., Chang, B. H. J., and Danesh, F. R. (2010). Identification of microRNA-93 as a novel regulator of vascular endothelial growth factor in hyperglycemic conditions. J Biol Chem 285, 23457–23465.

Lu, G., Dolgner, S. J., and Hall, T. M. T. (2009). Understanding and engineering RNA sequence specificity of PUF proteins. Curr Opin Struct Biol 19, 110–115.

Lu, G., and Hall, T. M. T. (2011). Alternate modes of cognate RNA recognition by human PUMILIO proteins. Structure 19, 361–367.

Lukong, K. E., Chang, K.-wei, Khandjian, E. W., and Richard, S. (2008). RNA-binding proteins in human genetic disease. Trends Genet 24, 416–425.

Lyng, M. B., Laenkholm, A.-V., Pallisgaard, N., and Ditzel, H. J. (2008). Identification of genes for normalization of real-time RT-PCR data in breast carcinomas. BMC Cancer 8, 20.

Mansfield, K. D., and Keene, J. D. (2009). The ribonome: a dominant force in co-ordinating gene expression. Biol Cell 101, 169–181.

Maragkakis, M., Reczko, M., Simossis, V. A., Alexiou, P., Papadopoulos, G. L., Dalamagas, T., Giannopoulos, G., Goumas, G., Koukis, E., Kourtis, K., et al. (2009). DIANA-microT web server: elucidating microRNA functions through target prediction. Nucleic Acids Res 37, W273–W276.

Matthaei, J. H., Jones, O. W., Martin, R. G., and Nirenberg, M. W. (1962). Characteristics and composition of RNA coding units. Proc Natl Acad Sci U S A 48, 666–677.

Mayer, G. (2009). The chemical biology of aptamers. Angew Chem Int Ed Engl 48, 2672–2689.

McArthur, K., Feng, B., Wu, Y., Chen, S., and Chakrabarti, S. (2011). MicroRNA-200b regulates vascular endothelial growth factor-mediated alterations in diabetic retinopathy. Diabetes 60, 1314–1323.

McManus, C. J., and Graveley, B. R. (2011). RNA structure and the mechanisms of alternative splicing. Curr Opin Genet Dev 21, 373–379.

Meiron, M., Anunu, R., Scheinman, E. J., Hashmueli, S., and Levi, B. Z. (2001). New isoforms of VEGF are translated from alternative initiation CUG codons located in its 5'UTR. Biochem Biophys Res Commun 282, 1053–1060.

Meng, F., Henson, R., Wehbe-Janek, H., Ghoshal, K., Jacob, S. T., and Patel, T. (2007). MicroRNA-21 regulates expression of the PTEN tumor suppressor gene in human hepatocellular cancer. Gastroenterology 133, 647–658.

Mesarovic, M. D., Sreenath, S. N., and Keene, J. D. (2004). Search for organising principles: understanding in systems biology. Syst Biol (Stevenage) 1, 19–27.

Millard, S. S., Vidal, A., Markus, M., and Koff, A. (2000). A U-rich element in the 5' untranslated region is necessary for the translation of p27 mRNA. Mol Cell Biol 20, 5947–5959.

Miller, M. A., and Olivas, W. M. (2011). Roles of Puf proteins in mRNA degradation and translation. Wiley Interdiscip Rev RNA 2, 471–492.

Milo, R., Shen-Orr, S., Itzkovitz, S., Kashtan, N., Chklovskii, D., and Alon, U. (2002). Network motifs: simple building blocks of complex networks. Science 298, 824–827.

Miquerol, L., Langille, B. L., and Nagy, A. (2000). Embryonic development is disrupted by modest increases in vascular endothelial growth factor gene expression. Development 127, 3941–3946.

Mohammad, M. M., Donti, T. R., Sebastian Yakisich, J., Smith, A. G., and Kapler, G. M. (2007). Tetrahymena ORC contains a ribosomal RNA fragment that participates in rDNA origin recognition. EMBO J 26, 5048–5060.

Morris, A. R., Mukherjee, N., and Keene, J. D. (2008). Ribonomic analysis of human Pum1 reveals cis-trans conservation across species despite evolution of diverse mRNA target sets. Mol Cell Biol 28, 4093–4103.

Morris, A. R., Mukherjee, N., and Keene, J. D. (2010). Systematic analysis of posttranscriptional gene expression. Wiley Interdiscip Rev Syst Biol Med 2, 162–180.

Mourelatos, Z., Dostie, J., Paushkin, S., Sharma, A., Charroux, B., Abel, L., Rappsilber, J., Mann, M., and Dreyfuss, G. (2002). miRNPs: a novel class of ribonucleoproteins containing numerous microRNAs. Genes Dev 16, 720–728.

Mukherjee, N., Corcoran, D. L., Nusbaum, J. D., Reid, D. W., Georgiev, S., Hafner, M.,

Ascano, M., Tuschl, T., Ohler, U., and Keene, J. D. (2011). Integrative Regulatory Mapping Indicates that the RNA-Binding Protein HuR Couples Pre-mRNA Processing and mRNA Stability. Mol Cell 43, 1–13.

Nelson, M. R., Luo, H., Vari, H. K., Cox, B. J., Simmonds, A. J., Krause, H. M., Lipshitz, H. D., and Smibert, C. A. (2007). A multiprotein complex that mediates translational enhancement in Drosophila. J Biol Chem 282, 34031–34038.

Niranjanakumari, S., Lasda, E., Brazas, R., and Garcia-Blanco, M. A. (2002). Reversible cross-linking combined with immunoprecipitation to study RNA-protein interactions in vivo. Methods 26, 182–190.

Nolde, M. J., Saka, N., Reinert, K. L., and Slack, F. J. (2007). The Caenorhabditis elegans pumilio homolog, puf-9, is required for the 3'UTR-mediated repression of the let-7 microRNA target gene, hbl-1. Dev Biol 305, 551–563.

Oliphant, A. R., Brandl, C. J., and Struhl, K. (1989). Defining the sequence specificity of DNA-binding proteins by selecting binding sites from random-sequence oligonucleotides: analysis of yeast GCN4 protein. Mol Cell Biol 9, 2944–2949.

Onesto, C., Berra, E., Grépin, R., and Pagès, G. (2004). Poly(A)-binding protein-interacting protein 2, a strong regulator of vascular endothelial growth factor mRNA. J Biol Chem 279, 34217–34226.

Pagano, M., Tam, S. W., Theodoras, A. M., Beer-Romero, P., Del Sal, G., Chau, V., Yew, P. R., Draetta, G. F., and Rolfe, M. (1995). Role of the ubiquitin-proteasome pathway in regulating abundance of the cyclin-dependent kinase inhibitor p27. Science 269, 682–685.

Palomero, T., Sulis, M. L., Cortina, M., Real, P. J., Barnes, K., Ciofani, M., Caparros, E., Buteau, J., Brown, K., Perkins, S. L., et al. (2007). Mutational loss of PTEN induces resistance to NOTCH1 inhibition in T-cell leukemia. Nat Med 13, 1203–1210.

Parisien, M., and Major, F. (2008). The MC-Fold and MC-Sym pipeline infers RNA structure from sequence data. Nature 452, 51–55.

Patzel, V., and Sczakiel, G. (1999). Length dependence of RNA-RNA annealing. J Mol Biol 294, 1127–1134.

Pineau, P., Volinia, S., McJunkin, K., Marchio, A., Battiston, C., Terris, B., Mazzaferro, V., Lowe, S. W., Croce, C. M., and Dejean, A. (2010). miR-221 overexpression contributes to liver tumorigenesis. Proc Natl Acad Sci U S A 107, 264–269.

Piqué, M., López, J. M., Foissac, S., Guigó, R., and Méndez, R. (2008). A combinatorial code for CPE-mediated translational control. Cell 132, 434–448.

Poltorak, Z., Cohen, T., Sivan, R., Kandelis, Y., Spira, G., Vlodavsky, I., Keshet, E., and Neufeld, G. (1997). VEGF145, a secreted vascular endothelial growth factor isoform that binds to extracellular matrix. J Biol Chem 272, 7151–7158.

Poy, M. N., Eliasson, L., Krutzfeldt, J., Kuwajima, S., Ma, X., Macdonald, P. E., Pfeffer, S.,

Tuschl, T., Rajewsky, N., Rorsman, P., et al. (2004). A pancreatic islet-specific microRNA regulates insulin secretion. Nature 432, 226–230.

Pullmann, R., Kim, H. H., Abdelmohsen, K., Lal, A., Martindale, J. L., Yang, X., and Gorospe, M. (2007). Analysis of turnover and translation regulatory RNA-binding protein expression through binding to cognate mRNAs. Mol Cell Biol 27, 6265–6278.

Pyle, A. M. (2010). The tertiary structure of group II introns: implications for biological function and evolution. Critic Rev Biochem Mol Biol 45, 215–232.

Quenault, T., Lithgow, T., and Traven, A. (2011). PUF proteins: repression, activation and mRNA localization. Trends Cell Biol 21, 104–112.

Ray, P. S., Jia, J., Yao, P., Majumder, M., Hatzoglou, M., and Fox, P. L. (2009). A stress-responsive RNA switch regulates VEGFA expression. Nature 457, 915–919.

Ray, P. S., and Fox, P. L. (2007). A post-transcriptional pathway represses monocyte VEGF-A expression and angiogenic activity. EMBO J 26, 3360–3372.

Rehmsmeier, M., Steffen, P., Hochsmann, M., and Giegerich, R. (2004). Fast and effective prediction of microRNA/target duplexes. RNA 10, 1507–1517.

Ren, B., Robert, F., Wyrick, J. J., Aparicio, O., Jennings, E. G., Simon, I., Zeitlinger, J., Schreiber, J., Hannett, N., Kanin, E., et al. (2000). Genome-wide location and function of DNA binding proteins. Science 290, 2306–2309.

Rouault, T. A., Hentze, M. W., Haile, D. J., Harford, J. B., and Klausner, R. D. (1989). The iron-responsive element binding protein: a method for the affinity purification of a regulatory RNA-binding protein. Proc Natl Acad Sci U S A 86, 5768–5772.

Ruby, S. W., and Abelson, J. (1988). An early hierarchic role of U1 small nuclear ribonucleoprotein in spliceosome assembly. Science 242, 1028–1035.

le Sage, C., Nagel, R., Egan, D. A., Schrier, M., Mesman, E., Mangiola, A., Anile, C., Maira, G., Mercatelli, N., Ciafrè, S. A., et al. (2007). Regulation of the p27(Kip1) tumor suppressor by miR-221 and miR-222 promotes cancer cell proliferation. EMBO J 26, 3699–3708.

Salazar, A. M., Silverman, E. J., Menon, K. P., and Zinn, K. (2010). Regulation of synaptic Pumilio function by an aggregation-prone domain. J Neurosci 30, 515–522.

Salven, P., Heikkilä, P., and Joensuu, H. (1997). Enhanced expression of vascular endothelial growth factor in metastatic melanoma. Br J Cancer 76, 930–934.

Sandberg, R., Neilson, J. R., Sarma, A., Sharp, P. a, and Burge, C. B. (2008). Proliferating cells express mRNAs with shortened 3' untranslated regions and fewer microRNA target sites. Science 320, 1643–1647.

Sauter, E. R., Nesbit, M., Watson, J. C., Klein-Szanto, A., Litwin, S., and Herlyn, M. (1999). Vascular endothelial growth factor is a marker of tumor invasion and metastasis in squamous cell carcinomas of the head and neck. Clin Cancer Res 5, 775–782.

Schmittgen, T. D., and Livak, K. J. (2008). Analyzing real-time PCR data by the comparative C(T) method. Nat Protoc 3, 1101–1108.

Schwartz, R. A., Bridges, T. M., Butani, A. K., and Ehrlich, A. (2008). Actinic keratosis: an occupational and environmental disorder. J Eur Acad Dermatol Venereol 22, 606–615.

Selbach, M., Schwanhäusser, B., Thierfelder, N., Fang, Z., Khanin, R., and Rajewsky, N. (2008). Widespread changes in protein synthesis induced by microRNAs. Nature 455, 58–63.

Senger, D. R., Galli, S. J., Dvorak, A. M., Perruzzi, C. A., Harvey, V. S., and Dvorak, H. F. (1983). Tumor cells secrete a vascular permeability factor that promotes accumulation of ascites fluid. Science 219, 983–985.

Shahbabian, K., and Chartrand, P. (2011). Control of cytoplasmic mRNA localization. Cell Mol Life Sci.

Shcherbakova, D. M., Sokolov, K. A., Zvereva, M. I., and Dontsova, O. A. (2009). Telomerase from yeast Saccharomyces cerevisiae is active in vitro as a monomer. Biochemistry (Mosc) 74, 749–755.

Shih, S. C., and Claffey, K. P. (1999). Regulation of human vascular endothelial growth factor mRNA stability in hypoxia by heterogeneous nuclear ribonucleoprotein L. J Biol Chem 274, 1359–1365.

Shyu, A.-B., Wilkinson, M. F., and van Hoof, A. (2008). Messenger RNA regulation: to translate or to degrade. EMBO J 27, 471–481.

Silva, I. J., Saramago, M., Dressaire, C., Domingues, S., Viegas, S. C., and Arraiano, C. M. Importance and key events of prokaryotic RNA decay: the ultimate fate of an RNA molecule. Wiley Interdiscip Rev RNA 2, 818–836.

Slingerland, J., and Pagano, M. (2000). Regulation of the cdk inhibitor p27 and its deregulation in cancer. J Cell Physiol 183, 10–17.

Slobodin, B., and Gerst, J. E. (2011). RaPID: an aptamer-based mRNA affinity purification technique for the identification of RNA and protein factors present in ribonucleoprotein complexes. Methods Mol Biol 714, 387–406.

Solomatin, S. V., Greenfeld, M., Chu, S., and Herschlag, D. (2010). Multiple native states reveal persistent ruggedness of an RNA folding landscape. Nature 463, 681–684.

Sonoda, J., and Wharton, R. P. (1999). Recruitment of Nanos to hunchback mRNA by Pumilio. Genes Dev 13, 2704–2712.

Sontheimer, E. J. (1994). Site-specific RNA crosslinking with 4-thiouridine. Mol Biol Rep 20, 35–44.

Spassov, D. S., and Jurecic, R. (2002). Cloning and comparative sequence analysis of PUM1 and PUM2 genes, human members of the Pumilio family of RNA-binding proteins. Gene 299, 195–204.

Spassov, D. S., and Jurecic, R. (2003). The PUF family of RNA-binding proteins: does evolutionarily conserved structure equal conserved function? IUBMB Life 55, 359–366.

Sponer, J., Leszczynski, J., and Hobza, P. (2002). Electronic properties, hydrogen bonding, stacking, and cation binding of DNA and RNA bases. Biopolymers 61, 3–31.

Srisawat, C., Goldstein, I. J., and Engelke, D. R. (2001). Sephadex-binding RNA ligands: rapid affinity purification of RNA from complex RNA mixtures. Nucleic Acids Res 29, E4.

Srisawat, C., Houser-Scott, F., Bertrand, E., Xiao, S., Singer, R. H., and Engelke, D. R. (2002). An active precursor in assembly of yeast nuclear ribonuclease P. RNA 8, 1348–1360.

Srisawat, C., and Engelke, D. R. (2002). RNA affinity tags for purification of RNAs and ribonucleoprotein complexes. Methods 26, 156–161.

Srisawat, C., and Engelke, D. R. (2001). Streptavidin aptamers: affinity tags for the study of RNAs and ribonucleoproteins. RNA 7, 632–641.

Steitz, J. A., and Vasudevan, S. (2009). miRNPs: versatile regulators of gene expression in vertebrate cells. Biochem Soc Trans 37, 931–935.

Sutton, D. H., Conn, G. L., Brown, T., and Lane, A. N. (1997). The dependence of DNase I activity on the conformation of oligodeoxynucleotides. Biochem J 321, Pt 2, 481–486.

Szabo, A., Perou, C. M., Karaca, M., Perreard, L., Palais, R., Quackenbush, J. F., and Bernard, P. S. (2004). Statistical modeling for selecting housekeeper genes. Genome Biol 5, R59.

Takizawa, P. A., and Vale, R. D. (2000). The myosin motor, Myo4p, binds Ash1 mRNA via the adapter protein, She3p. Proc Natl Acad Sci U S A 97, 5273–5278.

Tenenbaum, S. A., Lager, P. J., Carson, C. C., and Keene, J. D. (2002). Ribonomics: identifying mRNA subsets in mRNP complexes using antibodies to RNA-binding proteins and genomic arrays. Methods 26, 191–198.

Tenenbaum, S. a, Carson, C. C., Lager, P. J., and Keene, J. D. (2000). Identifying mRNA subsets in messenger ribonucleoprotein complexes by using cDNA arrays. Proc Natl Acad Sci U S A 97, 14085–14090.

Tereshko, V., Skripkin, E., and Patel, D. J. (2003). Encapsulating streptomycin within a small 40-mer RNA. Chem Biol 10, 175–187.

Thomas, M. G., Loschi, M., Desbats, M. A., and Boccaccio, G. L. (2011). RNA granules: the good, the bad and the ugly. Cell Signal 23, 324–334.

Thomas, P. D., Kejariwal, A., Guo, N., Mi, H., Campbell, M. J., Muruganujan, A., and Lazareva-Ulitsky, B. (2006). Applications for protein sequence-function evolution data: mRNA/protein expression analysis and coding SNP scoring tools. Nucleic Acids Res 34, W645–W650.

Tian, B., and Graber, J. H. (2011). Signals for pre-mRNA cleavage and polyadenylation. Wiley Interdiscip Rev RNA.

Topisirovic, I., Svitkin, Y. V., Sonenberg, N., and Shatkin, A. J. (2011). Cap and cap-binding proteins in the control of gene expression. Wiley Interdiscip Rev RNA 2, 277–298.

Tuerk, C., and Gold, L. (1990). Systematic evolution of ligands by exponential enrichment: RNA ligands to bacteriophage T4 DNA polymerase. Science 249, 505–510.

Tóth-Jakatics, R., Jimi, S., Takebayashi, S., and Kawamoto, N. (2000). Cutaneous malignant melanoma: correlation between neovascularization and peritumor accumulation of mast cells overexpressing vascular endothelial growth factor. Hum Pathol 31, 955–960.

Valencia-Burton, M., McCullough, R. M., Cantor, C. R., and Broude, N. E. (2007). RNA visualization in live bacterial cells using fluorescent protein complementation. Nat Methods 4, 421–427.

Vasudevan, S. (2011). Posttranscriptional Upregulation by MicroRNAs. Wiley Interdiscip Rev RNA.

Vasudevan, S., Tong, Y., and Steitz, J. A. (2007). Switching from repression to activation: microRNAs can up-regulate translation. Science 318, 1931–1934.

Vasudevan, S., and Steitz, J. A. (2007). AU-rich-element-mediated upregulation of translation by FXR1 and Argonaute 2. Cell 128, 1105–1118.

Vazquez-Pianzola, P., Urlaub, H., and Rivera-Pomar, R. (2005). Proteomic analysis of reaper 5' untranslated region-interacting factors isolated by tobramycin affinity-selection reveals a role for La antigen in reaper mRNA translation. Proteomics 5, 1645–1655.

Ventura, A., and Jacks, T. (2009). MicroRNAs and cancer: short RNAs go a long way. Cell 136, 586–591.

Verhelst, S. H. L., Michiels, P. J. A., van der Marel, G. A., van Boeckel, C. A. A., and van Boom, J. H. (2004). Surface plasmon resonance evaluation of various aminoglycoside-RNA hairpin interactions reveals low degree of selectivity. Chembiochem 5, 937–942.

Vessey, J. P., Schoderboeck, L., Gingl, E., Luzi, E., Riefler, J., Di Leva, F., Karra, D., Thomas, S., Kiebler, M. A., and Macchi, P. (2010). Mammalian Pumilio 2 regulates dendrite morphogenesis and synaptic function. Proc Natl Acad Sci U S A 107, 3222–3227.

Vessey, J. P., Vaccani, A., Xie, Y., Dahm, R., Karra, D., Kiebler, M. A., and Macchi, P. (2006). Dendritic localization of the translational repressor Pumilio 2 and its contribution to dendritic stress granules. J Neurosci 26, 6496–6508.

Viswanathan, S. R., Daley, G. Q., and Gregory, R. I. (2008). Selective blockade of microRNA processing by Lin28. Science 320, 97–100.

Vumbaca, F., Phoenix, K. N., Rodriguez-Pinto, D., Han, D. K., and Claffey, K. P. (2008). Double-stranded RNA-binding protein regulates vascular endothelial growth factor mRNA stability, translation, and breast cancer angiogenesis. Mol Cell Biol 28, 772–783.

Walker, S. C., Good, P. D., Gipson, T. A., and Engelke, D. R. (2011). The dual use of RNA aptamer sequences for affinity purification and localization studies of RNAs and RNA-protein complexes. Methods Mol Biol 714, 423–444.

Wallace, S. T., and Schroeder, R. (1998). In vitro selection and characterization of streptomycin-binding RNAs: recognition discrimination between antibiotics. RNA 4, 112–123.

Wang, X. (2008). miRDB: a microRNA target prediction and functional annotation database with a wiki interface. RNA 14, 1012–1017.

Wang, X., McLachlan, J., Zamore, P. D., and Hall, T. M. T. (2002). Modular recognition of RNA by a human pumilio-homology domain. Cell 110, 501–512.

Wang, X., Zamore, P. D., and Hall, T. M. (2001). Crystal structure of a Pumilio homology domain. Mol Cell 7, 855–865.

Wang, Y., Killian, J., Hamasaki, K., and Rando, R. R. (1996). RNA molecules that specifically and stoichiometrically bind aminoglycoside antibiotics with high affinities. Biochemistry 35, 12338–12346.

Wang, Y., and Rando, R. R. (1995). Specific binding of aminoglycoside antibiotics to RNA. Chem Biol 2, 281–290.

Wang, Z., Gerstein, M., and Snyder, M. (2009). RNA-Seq: a revolutionary tool for transcriptomics. Nat Rev Genet 10, 57–63.

Wang, Z., Kayikci, M., Briese, M., Zarnack, K., Luscombe, N. M., Rot, G., Zupan, B., Curk, T., and Ule, J. (2010). iCLIP predicts the dual splicing effects of TIA-RNA interactions. PLoS Biol 8, e1000530.

Wapinski, O., and Chang, H. Y. (2011). Long noncoding RNAs and human disease. Trends Cell Biol 21, 354–361.

Weinlich, S., Hüttelmaier, S., Schierhorn, A., Behrens, S.-E., Ostareck-Lederer, A., and Ostareck, D. H. (2009). IGF2BP1 enhances HCV IRES-mediated translation initiation via the 3'UTR. RNA 15, 1528–1542.

Welting, T. J. M., Mattijssen, S., Peters, F. M. A., van Doorn, N. L., Dekkers, L., van Venrooij, W. J., Heus, H. A., Bonafé, L., and Pruijn, G. J. M. (2008). Cartilage-hair hypoplasia-associated mutations in the RNase MRP P3 domain affect RNA folding and ribonucleoprotein assembly. Biochim Biophy Acta 1783, 455–466.

Wepf, A., Glatter, T., Schmidt, A., Aebersold, R., and Gstaiger, M. (2009). Quantitative interaction proteomics using mass spectrometry. Nat Methods 6, 203–205.

White, E. K., Moore-Jarrett, T., and Ruley, H. E. (2001). PUM2, a novel murine puf protein, and its consensus RNA-binding site. RNA 7, 1855–1866.

Whittle, C., Gillespie, K., Harrison, R., Mathieson, P. W., and Harper, S. J. (1999). Heterogeneous vascular endothelial growth factor (VEGF) isoform mRNA and receptor mRNA expression in human glomeruli, and the identification of VEGF148 mRNA, a novel truncated splice variant. Clin Sci (Lond) 97, 303–312.

Wickens, M., Bernstein, D. S., Kimble, J., and Parker, R. (2002). A PUF family portrait: 3'UTR regulation as a way of life. Trends Genet 18, 150–157.

Williamson, J. R. (2000). Induced fit in RNA-protein recognition. Nat Struct Biol 7, 834–837.

Windbichler, N., and Schroeder, R. (2006). Isolation of specific RNA-binding proteins using the streptomycin-binding RNA aptamer. Nat Protoc 1, 637–640.

Winter, J., Jung, S., Keller, S., Gregory, R. I., and Diederichs, S. (2009). Many roads to maturity: microRNA biogenesis pathways and their regulation. Nat Cell Biol 11, 228–234.

Wreden, C., Verrotti, A. C., Schisa, J. A., Lieberfarb, M. E., and Strickland, S. (1997). Nanos and pumilio establish embryonic polarity in Drosophila by promoting posterior deadenylation of hunchback mRNA. Development 124, 3015–3023.

Wu, H., Zhu, S., and Mo, Y.-Y. (2009). Suppression of cell growth and invasion by miR-205 in breast cancer. Cell Res 19, 439–448.

Wu, L., and Belasco, J. G. (2005). Micro-RNA regulation of the mammalian lin-28 gene during neuronal differentiation of embryonal carcinoma cells. Mol Cell Biol 25, 9198–9208.

Xiao, S., Day-Storms, J. J., Srisawat, C., Fierke, C. A., and Engelke, D. R. (2005). Characterization of conserved sequence elements in eukaryotic RNase P RNA reveals roles in holoenzyme assembly and tRNA processing. RNA 11, 885–896.

Yakovchuk, P., Protozanova, E., and Frank-Kamenetskii, M. D. (2006). Base-stacking and base-pairing contributions into thermal stability of the DNA double helix. Nucleic Acids Res 34, 564–574.

Yang, G., Fu, H., Zhang, J., Lu, X., Yu, F., Jin, L., Bai, L., Huang, B., Shen, L., Feng, Y., et al. (2010). RNA-binding protein quaking, a critical regulator of colon epithelial differentiation and a suppressor of colon cancer. Gastroenterology 138, 231–240.e1 c5.

Yang, G., Lu, X., Wang, L., Bian, Y., Fu, H., Wei, M., Pu, J., Jin, L., Yao, L., and Lu, Z. (2011). E2F1 and RNA binding protein QKI comprise a negative feedback in the cell cycle regulation. Cell Cycle 10, 2703–2713.

Yang, H., Kong, W., He, L., Zhao, J.-J., O'Donnell, J. D., Wang, J., Wenham, R. M., Coppola, D., Kruk, P. A., Nicosia, S. V., et al. (2008). MicroRNA expression profiling in human ovarian cancer: miR-214 induces cell survival and cisplatin resistance by targeting PTEN. Cancer Res 68, 425–433.

Ye, W., Lv, Q., Wong, C.-K. A., Hu, S., Fu, C., Hua, Z., Cai, G., Li, G., Yang, B. B., and Zhang, Y. (2008). The effect of central loops in miRNA:MRE duplexes on the efficiency of miRNA-mediated gene regulation. PLoS One 3, e1719.

Yigit, E., Batista, P. J., Bei, Y., Pang, K. M., Chen, C.-C. G., Tolia, N. H., Joshua-Tor, L., Mitani, S., Simard, M. J., and Mello, C. C. (2006). Analysis of the C. elegans Argonaute family reveals that distinct Argonautes act sequentially during RNAi. Cell 127, 747–757.

Zhang, C., and Darnell, R. B. (2011). Mapping in vivo protein-RNA interactions at single-nucleotide resolution from HITS-CLIP data. Nat Biotechnol 29, 607–614.

Zhang, C.-D., Pan, M.-H., Tan, J., Li, F.-F., Zhang, J., Wang, T.-T., and Lu, C. (2011). Characteristics and evolution of the PUF gene family in Bombyx mori and 27 other species. Mol Biol Rep.

Zhao, Z., Chang, F. C., and Furneaux, H. M. (2000). The identification of an endonuclease that cleaves within an HuR binding site in mRNA. Nucleic Acids Res 28, 2695–2701.

Zheng, Y., and Miskimins, W. K. (2011). CUG-binding protein represses translation of p27Kip1 mRNA through its internal ribosomal entry site. RNA Biol 8, 365–371.

Zhou, S., Gu, L., He, J., Zhang, H., and Zhou, M. (2011). MDM2 regulates VEGF mRNA stabilization in hypoxia. Mol Cell Biol, [Epub ahead of print].

Zhou, Z., Licklider, L. J., Gygi, S. P., and Reed, R. (2002). Comprehensive proteomic analysis of the human spliceosome. Nature 419, 182–185.

Zhou, Z., and Reed, R. (2003). Purification of functional RNA-protein complexes using MS2-MBP. Curr Protoc Mol Biol Chapter 27, Unit 27.3.

Ziegeler, G., Ming, J., Koseki, J. C., Sevinc, S., Chen, T., Ergun, S., Qin, X., and Aktas, B. H. (2010). Embryonic lethal abnormal vision-like HuR-dependent mRNA stability regulates post-transcriptional expression of cyclin-dependent kinase inhibitor p27Kip1. J Biol Chem 285, 15408–15419.

Zielinski, J., Kilk, K., Peritz, T., Kannanayakal, T., Miyashiro, K. Y., Eiríksdóttir, E., Jochems, J., Langel, U., and Eberwine, J. (2006). In vivo identification of ribonucleoprotein-RNA interactions. Proc Natl Acad Sci U S A 103, 1557–1562.

9 Acknowledgments

This dissertation would not have been possible without the creativity and support of my supervisor André Gerber. His enthusiasm for this fascinating topic has been truly inspirational. I would like to thank you very much and wish you all the best for your future career!

I am deeply indebted and thankful to Michael Detmar for his supervision and continuous support, both personally and scientifically. I have greatly enjoyed my time in your lab and appreciate the scientific freedom and motivating environment that you provided for everyone through your hard and diligent work.

I am very grateful to Jonathan Hall for his input and support as a collaborator, member of my thesis committee and co-referee for this thesis. Your dedication and your sharp and inquisitive mind have been a great motivation, and your occasional 'words of wisdom' will not be forgotten.

I would like to thank all the past and present members of the Gerber, Detmar, Halin and Hall labs for the outstanding atmosphere, scientific discussions, and valuable diversion in the form of (some very memorable!) extracurricular events and excursions. I am especially thankful for the "behind-the-scences" work of Cornelius Fischer and Susanne Holliger. To Lucy: May he become the first rock star with a serious chocolate milk habit! To Acy: Türkçe dersleri için sağ ol. Merhaba akülü mämälär [sic!] onbir matkap ebediyen.

My heartfelt thanks and best wishes also go out to my students Fabi, Alex, Kasia, and particularly the "geezer" Felix, who became a dear friend.

Acknowledgments

Importantly, I would like to express my gratitude to all the other people who have helped me with experiments, reagents or advice: Piotr Dziunycz, Günther Hofbauer, Michael Hengartner, Boris Günnewig, Julian Zagalak, Andreas Frei, Bernd Wollscheid, Alexander Wepf, Matthias Gstaiger, Frank Wippich, Christoph Rösli, Gunther Meister, Martijn Kedde, Reuven Agami, and Susanna Bachmann. It has been a pleasure working with each and everyone of you, and your contributions and insights are truly appreciated.

Gladly I would like to thank Nina, Asli and Messer for being around and making me feel at home during the last ten years or so. Who knows what the future holds, but I would be more than happy if it were not too far away from you.

Kedves Fakopács: Nagyon köszönöm a sok szeretetet és a támogatást. Nélküled nem lettem volna képes erre. Két fiú, egy lány, és ötven+ év... Szeretlek.

Es ist mir die grösste Freude meinen Eltern Marita und Helmut, sowie meinen Grosseltern Pik, Elfriede, Willi und Gottfried für Ihre Hingabe, Ihre Fröhlichkeit, Ihr Vertrauen und Ihre bedingungslose Unterstützung zu danken. Ohne sie hätte ich das nie geschafft. Ihr seid die Besten!

10 Abbreviations

2'-O-Me	2'-O-methylated (or 2'-methoxy)
ARE	AU-rich element
DMEM	Dulbecco's modified Eagle's medium
DNA	deoxyribonucleic acid
eGFP	enhanced green fluorescent protein
ELAVL1/HuR	ELAV (embryonic lethal, abnormal vision, Drosophila)-like 1 (Hu antigen R)
ELISA	enzyme-linked immunosorbent assay
EtOH	ethanol
FRT	flippase recognition target
G	guanine
GO	gene ontology
GRN	gene regulatory network
HITS-CLIP	high-throughput sequencing crosslinking and immunopurification
iCLIP	individual nucleotide resolution crosslinking and immunopurification
hnRNP	heterogeneous nuclear ribonucleoprotein
HPLC	high-performance liquid chromatography
IRES	internal ribosome entry site
K_d	dissociation constant
LC-MS	liquid chromatography-mass spectrometry
LNA	locked nucleic acid
MFE	minimum free energy
miRNA	microRNA
MW	molecular weight
nt	nucleotide
PAR CLIP	photoactivatable ribonucleoside enhanced crosslinking and immunopurification
PBS	phosphate-buffered saline
PNA	peptide nucleic acids
PTGR	post-transcriptional gene regulation
PUF	pumilio/Fem-3-binding
RBP	RNA-binding protein

Abbreviations

RIP-Chip	RNA-binding protein immunopurification-microarray
RLU	relative luciferase units
RNA	ribonucleic acid
RNase P	ribonuclease P
RNP	ribonucleoprotein
RP-HPLC	reversed phase HPLC
rRNA	ribosomal RNA
RT	room temperature (22°C)
S.D.	standard deviation
S.E.M.	standard error of the mean
SELEX	systematic evolution of ligands by exponential enrichment
snoRNA	small nucleolar RNA
snRNA	small nuclear RNA
snRNP	small nuclear ribonucleoprotein
SV40	simian virus 40
UTR	untranslated region
UV	ultraviolet
wt	wild type

i want morebooks!

Buy your books fast and straightforward online - at one of world's fastest growing online book stores! Environmentally sound due to Print-on-Demand technologies.

Buy your books online at
www.get-morebooks.com

Kaufen Sie Ihre Bücher schnell und unkompliziert online – auf einer der am schnellsten wachsenden Buchhandelsplattformen weltweit! Dank Print-On-Demand umwelt- und ressourcenschonend produziert.

Bücher schneller online kaufen
www.morebooks.de

VDM Verlagsservicegesellschaft mbH
Heinrich-Böcking-Str. 6-8   Telefon: +49 681 3720 174   info@vdm-vsg.de
D - 66121 Saarbrücken   Telefax: +49 681 3720 1749   www.vdm-vsg.de

Printed by Books on Demand GmbH, Norderstedt / Germany